Geology of Southeast Alaska

Rock and Ice in Motion

Geology of Southeast Alaska
Rock and Ice in Motion

Harold H. Stowell

University of Alaska Press
Fairbanks

© 2006 University of Alaska Press

P.O. Box 756240
Fairbanks, AK 99775-6240

Printed in China

This publication was printed on paper that meets the minimum
requirements for ANSI/NISO x39.48-1992 (Permanence of Paper).

ISBN 13: 978-1-889963-81-5
ISBN 10: 1-889963-81-X

Library of Congress Cataloging-in-Publication Data
Stowell, Harold Hilton.
Geology of southeast Alaska : rock and ice in motion / Harold H. Stowell.
    p. cm.
Includes bibliographical references.
ISBN-13: 978-1-889963-81-5 (pbk. : alk. paper)
ISBN-10: 1-889963-81-X (pbk. : alk. paper)
1. Geology—Alaska, Southeast. 2. Glaciers—Alaska, Southeast.
3. Plate tectonics—Alaska, Southeast. 4. Climate—Alaska, Southeast. I. Title.
QE84.S68S76 2006
557.982—dc22
                            2005027066

All illustrations by Harold H. Stowell unless otherwise noted.
Cover and interior text design by Dixon J. Jones, UAF Rasmuson Library Graphics
Book production by Sue Mitchell

Cover image: the Fairweather Range; the terminus of Brady Glacier is visible at the
base of the mountains.

Back cover images: (top) sea arch on Kuiu Island; (center) infrared aerial photograph of Margerie, Ferris, and Grand Pacific glaciers; (bottom) glacial grooves cut
into the Coast Plutonic Complex sill along the south shore of Tracy Arm. Sun and
glacial silt produce the intense blue-green water color.

*To those who contribute
to understanding the geology
and to preserving the unique environment
of Southeast Alaska and especially to Kim Ouderkirk,
who provided moral and physical support
throughout the research and writing of this book.*

Table of Contents

Acknowledgments

The geologic story presented here and the projects that I have undertaken in Southeast Alaska rely on research that many people have done over the last one hundred years. The U.S. Geological Survey (USGS) produced much of the geological mapping, which provides the basis for later, more detailed studies. One of the earliest comprehensive works, published in 1929, is Buddington and Chapin's *Geology and Mineral Deposits of Southeastern Alaska*. The USGS has completed many important maps and studies in Southeast Alaska since 1929, many of which can be attributed to the mapping efforts of Henry Berg and David Brew. Finally, the 1992 compilation entitled *Geologic Map of Southeastern Alaska* by Gehrels and Berg provides a wonderful starting point for understanding the geology.

My research and interest in Southeast Alaska have benefited from many helpful colleagues and friends. Kim Ouderkirk provided support throughout my research and writing. Helpful individuals have provided research and job opportunities, transport to remote areas, hot food and showers after many days in the cold rain, and insight into geological complexities. Some of the important people who have helped me with my work are David Brew, B. F. Buie, Lincoln Hollister, Russ Ives, Dan Krehbeil, Aral Loken, Jack O'Hara, Doug Scudder, and Ron Short. I thank my friends at Lindblad Expeditions for making many of my travels in Southeast

Alaska possible. Finally, this book has greatly benefited from the tireless encouragement and editing of Jennifer Collier, and from helpful reviews by Greg Streveler and Cathy Connor.

Introduction

Southeast Alaska is a geologically active and fascinating part of the world. Current geological activity is manifest in earthquakes, glacial movement, and the less perceptible uplift and *erosion* of the earth's *crust* (terms in *italics* are defined in the glossary). Records of ancient geological events are well preserved in the fascinating rocks exposed in the glacially carved landscape. These rocks have been used to reconstruct a record of *volcano* activity, *sedimentation,* and deep crustal processes. These processes have occurred mostly since the beginning of the Mesozoic Era (more than 200 million years ago) of earth's history.

Towering mountains throughout the region continue to grow by *tectonic* uplift, a seemingly slow process that occurs at rates of less than 0.5 to 1.5 inches (1.3 to 3.8 cm) per year (past rates may have been somewhat higher). Large-scale erosion by the relentless flow of *glaciers* and rivers, which occurs at a rate similar to uplift, has removed most of the original mountains that began to grow more than 90 million years ago. These active processes of mountain building and erosion in Southeast Alaska have created the vast network of waterways that provide an ideal route for viewing the diverse geology and biology of this region.

I wrote this book to guide travelers through the spectacular scenery and fascinating geology of Southeast Alaska. It provides an introduction to the complex, glacially carved geography of the region; to the processes and geology associated with *plate tectonics, faults*, earthquakes, mountain building, and *exotic terranes*; and to the geology of the Coast Mountains. The final chapter, "Regional Geology," describes and interprets specific geological features found in some of the most visited and spectacular locations in Southeast Alaska.

CHAPTER 1

Geography of Southeast Alaska

S outheast Alaska was shaped by the advance and retreat of glacial ice, which at various times covered most of the region. Thick ice sheets carved mountainous islands and a high ridge of coastal mountains on the mainland (Figure 1), during the last seven million years of earth history. Between 1.8 million and ten thousand years ago, a period geologists call the Pleistocene Epoch, extensive *glaciation* occurred in much of northern North America. The most recent retreat of a continental ice sheet began about twenty thousand years ago. The retreat of the ice that covered most of Southeast Alaska as recently as fourteen thousand years ago left dramatic, mountainous islands that provide stunning evidence for the power of ice to carve solid rock. Today all that remains of the vast continental ice sheet are relatively small alpine glaciers.

The numerous islands of Southeast Alaska were named the Alexander Archipelago in honor of Tsar Alexander II. The Russian and English explorers named the four largest islands Prince of Wales, Baranof, Chichagof, and Admiralty. Surrounding them are about eleven hundred smaller islands, some of which are themselves substantial in size (Figure 1). The indigenous Tlingit and Haida peoples had their own names for these islands and the

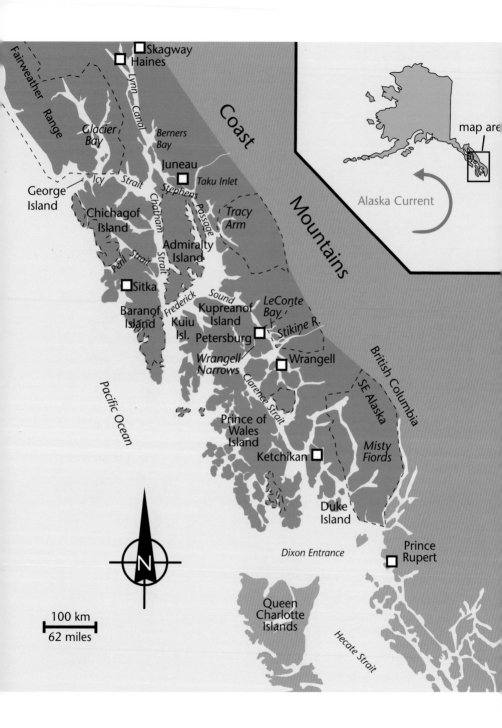

waterways that connect them long before the arrival of Europeans. For example, the Tlingit name for Angoon on Admiralty Island is Kootznoowoo, which means "brown bear den," or is sometimes interpreted as "fortress of the bears." All of the islands have complex coastlines of bays, inlets, and river estuaries. The result is a rugged, generally tree-covered shoreline (Figure 2) of more than 11,000 miles (17,700 km).

The Alexander Archipelago is bordered to the northwest and northeast by the tall, ice-capped Fairweather Range and Coast Mountains, respectively. These mainland ranges, which contain most of the glaciers in Southeast Alaska, form the boundary between Alaska and British Columbia, Canada. Many of the peaks in these mountains are more than 6,000 feet (1,830 m) above sea level. The highest mountains, in the Fairweather Range west of Glacier Bay, exceed 15,000 feet (4,570 m) in elevation (Figure 3).

Southeast Alaska has a moderate maritime climate, in spite of its northerly latitude of 54 to 58° north. Mean winter temperatures are around 30°F (–1°C), lower inland and to the north. Mean summer temperatures are around 60°F (16°C). This mild weather is largely due to the Alaska and Japan currents that transport tropical and subtropical ocean water north into the Gulf of Alaska (Figure 1). Still, ocean water temperatures are a bit chilly for swimming (40 to 50°F or 4 to 10°C during the summer).

Westerly winds traveling across the Gulf of Alaska carry warm, moist air onshore. These onshore winds create mild temperatures and copious amounts of coastal precipitation. When humid air from

◄ **FIGURE 1.** Geography of Southeast Alaska and adjacent British Columbia. Dashed lines enclose the largest preserved areas of wild lands, including Glacier Bay National Park and Preserve, Admiralty Island National Monument, Tracy Arm–Fords Terror Wilderness, and Misty Fiords National Monument. Inset map shows the Gulf of Alaska and the approximate path of warm water flowing northward in the Japan-Alaska current.

FIGURE 2. The heavily forested and rugged shoreline shrouded by fog along Peril Strait, Chichagof, and Baranof islands.

FIGURE 3. High peaks of the Fairweather Range form the skyline with Tsirku Glacier, British Columbia, in the foreground, looking southwest toward Glacier Bay. These high peaks accumulate vast amounts of snow that feed numerous icefields and glaciers.

the Pacific hits the mountains, it rises, cools, and drops moisture as rain or snow. Therefore, the western sides and tops of mountains are generally very wet, while the eastern sides of the mountains receive significantly less precipitation. For example, Little Port Walter on the southern end of Baranof Island, which has only one low, narrow ridge of mountains between it and the Pacific, receives an annual average of more than 200 inches (508 cm). The city of Juneau, on the mainland, is sheltered by several mountain ranges and receives an average of "only" 70 inches (180 cm) of precipitation each year.

Moist air flowing over the mountains creates not only rain, snow, and glaciers, but also the lush, temperate rainforest that covers most of the lower elevations in Southeast Alaska (Figure 4). These rainforests, most of which are within the boundaries of Tongass National Forest, are dominated by four species of large evergreen trees: Sitka spruce, western hemlock, western red cedar

FIGURE 4. Lush estuarine meadow flanked by alders and Sitka spruce along the shoreline of Hanus Bay, Baranof Island.

(in the southern half of Southeast Alaska), and Alaska yellow cedar. The dense forest is broken by coastal meadows of grasses and shrubs, extensive open bogs known as muskeg, and numerous scars left by landslides and avalanches. Salt-tolerant plants and a thick line of Sitka alder line much of the coastline. Lesser numbers of other deciduous trees such as black cottonwood, red alder, and shrubby Sitka mountain ash can be found in favorable habitat. Tree line varies from less than 1,500 feet (457 m) to about 3,500 feet (1,067 m) above sea level, depending on average snowfall and winds. Alpine areas vary from lush, herbaceous meadows and heaths with stunted mountain hemlock and subalpine fir to sheer rock peaks and *icefields*.

CHAPTER 2

Glaciers, Ice Ages, and Global Change

T he relatively warm water and moist air that make the climate of Southeast Alaska so mild also feed mountain glaciers along the northeast Pacific coast from the Cascades in Washington state to Southcentral Alaska. Today, icefields and glaciers are largely confined to areas in or near high mountains. However, geologists have determined that in the recent past these glaciers and others throughout the world were far more extensive than they are today.

Studies of glacial sediments, glacial erosion, and *paleoclimate* indicate that numerous *ice ages* have occurred in the history of the earth. These ice ages were times during which ice cover was far more extensive than it is today. During the Pleistocene, continental ice sheets periodically advanced across North America. Although the record is incomplete, ice probably covered large parts of southern coastal Alaska even before the Pleistocene. At least seven major ice ages have occurred in the last seven hundred thousand years,[1] with the last one ending about fourteen thousand years ago. Although there were immense glaciers that covered much of North America during this last ice age, temperatures were probably only 4 to 7°F (2.5 to 4°C) cooler than those of the present.[2] Even historically, the earth has been significantly cooler

7

than at present, for example during the Little Ice Age from AD 1400 to 1750.

Ice ages may have been caused by several different mechanisms:

1. Global changes in the absorption of thermal energy from the sun, caused by variations in the earth's orbit known as *Milankovitch cycles*.
2. Changes in the chemistry of the earth's atmosphere (*greenhouse effect*) that affect heat retention.
3. Changes in energy from the sun itself.
4. Movement of continents toward and away from the earth's poles. (However, movement of continents is very slow and could not be responsible for Pleistocene ice ages.)

The first explanation, Milankovitch cycles, is probably the most widely accepted cause for Pleistocene ice ages. Changes in earth's orbital shape, tilt of the axis of rotation, and wobble of the rotational axis vary in a cyclic fashion in periods of 100,000, 41,000, and 23,000 years. Paleoclimate models of the combined cycles predict changes in absorption of solar energy that match the paleoclimate record reasonably well.[3] Discounting human impacts on climate, these cycles suggest that we are near the middle of an interglacial (between ice ages) period, and that the climate should cool in the next few thousand years.

The second explanation, directly related to the greenhouse effect, is also likely to have played a significant role in Pleistocene ice ages. When solar energy hits the earth, some of it is reflected back into space and some of it is absorbed at the surface. Energy that is absorbed at the surface is emitted as infrared energy that is absorbed and re-emitted by gases in the atmosphere. Carbon dioxide, methane, and water in the atmosphere are particularly effective at absorbing infrared energy, causing a heating of the lower atmosphere. This heat retention has made the planet habitable to plants and animals and is known as the greenhouse effect. However, abundant geological data suggest that the earth's atmospheric chemistry changes dramatically with time. Lower concentrations of the greenhouse gases would reduce the retention of solar energy and lower temperatures. Today it is a great concern that large increases in greenhouse gases, for example as a result of burning fossil fuels and vegetation, may cause a significant temperature increase that could be devastating to many forms of life.

The third explanation for the onset of ice ages, changes in the sun's energy, is supported by observations of sunspots, growth rings in long-lived trees, and ancient ice chemistry. These data have been used to suggest that solar brightness could have caused the Little Ice Age. If changes in solar output during the Pleistocene were as dramatic as those inferred from the Little Ice Age data, they could have been responsible for significant climate changes.[4] However, we have no unambiguous data for solar energy output prior to 1978 when satellite measurements began. Therefore, no quantitative correlations between ice ages and solar energy have been made and the role of solar output in producing ice ages remains somewhat speculative.

Many methods are being developed and used to study climate history. Trapped air is one phenomenon that provides intriguing clues into atmospheric chemistry and climate history. As snow crystals fall on glaciers, accumulate, and pack into ice, they trap air from the atmosphere, some of which can remain in glacial ice for thousands of years. Trapped air in ice cores provides a record of atmosphere composition, and the oxygen composition (ratio of types of oxygen atoms or *isotopes*) in the ice provides an estimate of global ice volumes and temperatures (see inset "Estimating Global Ice Volume and Temperature"). The age of trapped air can be determined because the snow accumulates in seasonal layers, formed in cooler months. These annual layers can be counted like tree rings. Ages determined in this fashion can then be confirmed by comparison to ages determined from chemical or isotopic data.

Large long-lived continental ice sheets thus preserve a long climate record. Recent studies of continental ice sheets in Greenland have allowed partial reconstruction of atmospheric chemistry and ice volumes during the last 200,000 years. Ice from the Antarctic has recently yielded a record extending back as far as 740,000 years.[5] All of these ice core studies show a strong negative correlation between ice volume and carbon dioxide concentrations in the trapped atmospheric gases. Ice volumes were greatest when carbon dioxide concentrations were lowest. The cyclic variations match the Milankovitch cycles, which cause variable absorption of solar energy. However, it is not clear whether atmospheric carbon dioxide causes or results from warmer temperatures in part because of uncertainties in the timing that preclude determination of which occurs first in each cycle. In conclusion, scientists are only beginning to understand the complex interplay between biology, geology, and astronomy. Therefore, we cannot yet be certain about predicting the causes for past climate variation. Regardless, these data, combined with other scientific advances, are likely to provide significant advances in understanding of paleoclimate.

Estimating Global Ice Volume and Temperature

Stable isotopes of common elements can provide important information about the earth's energy budget and interactions between the atmosphere, oceans, and rocks. For example, two isotopes of oxygen—oxygen-16 and oxygen-18—provide information about global evaporation from the oceans and ocean temperatures. These oxygen isotopes are stable (do not decay) and occur naturally in water, air, and rocks. Evaporation of water preferentially concentrates the lighter oxygen-16 isotope in atmospheric water vapor. Weather systems transport the oxygen-16-rich water vapor toward the poles where some of it precipitates as snow. This snow is added to ice sheets like those on Antarctica and Greenland.

The difference in isotopic composition of water in ice sheets and the ocean water depends on the volume of ice, which in turn depends on global climate. Isotopic ratios of water in the ocean are recorded in the shells of marine micro-organisms (for example, *foraminifera*). Concentrations of the heavier oxygen isotope, oxygen-18, in these shells may indicate higher rates of evaporation and thus warmer global temperatures. However, the ratio depends both on local water temperature as the shell grows and on the isotopic composition of the ocean. Therefore, we cannot calculate a certain temperature or ice volume for the past without additional information. Fortunately, we can estimate surface ocean temperatures from the species of fossil micro-organisms that live there (for example, *radiolaria*) because these organisms can only survive in very narrow ranges of temperature.

Another strategy is to assume that very deep ocean water temperatures have not changed significantly. If we assume no deep ocean temperature changes, the ocean isotopic composition can be reconstructed from the foraminifera that live on the ocean floor. Isotopic ratios of water vapor in the atmosphere are recorded in snow and can be measured in ice cores. With records of oxygen isotope ratios in ice cores and fossil shells, and estimated ocean temperatures, a relative record of ice volume can be constructed. Somewhat reassuringly, estimates for large ice volumes closely match estimates for low ocean temperatures during the last 1.5 million years.[6]

Clearly, changes in climate have had and will continue to have a tremendous impact on the North. Presently, ice covers about 35,156 square miles (90,000 km^2) in Alaska and adjacent provinces

of Canada. This represents about 13 percent of the earth's mountain glacier area.[7] Although this percentage may seem fairly low,
its significance with respect to global change is probably quite
large. Arendt and others[8] recently estimated that melting of Alaska
glaciers is rapid, has recently accelerated, and contributes nearly 8
cubic miles (33 km^3) of water to the oceans each year. This water is
the largest glacial contribution to rising sea level—approximately
double the estimated contribution from melting of the Greenland
ice sheet—and is sufficient to produce a global rise in sea level of
about 0.01 inch/year (0.25 mm/yr).

Glaciers, Icefields, and Icebergs

Large snowfields, glaciers, and icefields can be seen throughout Southeast Alaska. Although from a distance they may look like
large, indistinguishable fields of snow, there are some important
distinctions between them. Snowfields are large areas of partially
recrystallized snow (not solid ice) that may persist throughout the
year (perennial). Glaciers, on the other hand, are areas of ice that
show evidence of past or present movement. Snow becomes a glacier when it accumulates into layers that become thick enough
(about 150 feet or 46 m) to cause the snow to recrystallize into
coarse interlocking ice crystals and move by *plastic flow*. Icefields
are large areas of ice accumulation that feed multiple glaciers.
Glaciers are categorized as alpine (or valley) or continental. Alpine
glaciers are confined to valleys and fed by high snowfall in mountainous regions. Continental glaciers cover large areas and are not
confined to valleys. Today continental glaciers and similar ice
caps are only found in Greenland, Antarctica, and polar islands.
Movement of all glacial ice is controlled by gravity so it always
flows downhill, away from mountains or away from the thickest
and highest parts of continental glaciers.

Icefields are large areas of gently sloping ice that typically feed several glaciers flowing radially outward from them. These areas of ice tend to form high in the mountains where temperatures are low and vast amounts of snow falls each year. In Southeast Alaska, the Stikine Icefield east of Petersburg feeds the LeConte Glacier, the Juneau Icefield northeast of Juneau feeds Mendenhall and Taku glaciers, and icefields north and east of Glacier Bay feed the Margerie, Grand Pacific, and Johns Hopkins glaciers around Glacier Bay.

Glacial ice that flows into the ocean results in spectacular tidewater glaciers such as those seen in Glacier Bay and Tracy Arm (Figure 5). Movement of the ice toward the ocean, interaction with waves and currents, and melting of ice by the saltwater are

FIGURE 5. The terminus of Sawyer Glacier nestled between steep mountains. Sawyer Glacier is retreating rapidly, and the line bounding lightly colored bare rock above the ice is evidence for the recent extent of ice. Pieces of ice frequently calve or break off the glacier, creating a spectacular scene and an ideal place for harbor seals to pup during May and June.

all forces that break off pieces of the glacier. These pieces of ice, known as *icebergs*, float because the density of the ice (about 56 pounds per cubic foot or 0.9 gm/cm^3) is less than that of ocean water (62 pounds per cubic foot or 1 gm/cm^3). Mariners have specific terms to describe the size of the floating glacial ice.[9] The largest are icebergs and rise more than 16.7 feet (5 m) above water. Medium-sized ice pieces called *bergy bits* extend between 3.3 and 16.7 feet (1 to 5 m) above water and are less than 33 feet (10 m) across. The smallest pieces are called *growlers* and extend less than 3.3 feet (1 m) above water and are generally less than 20 feet (6 m) across.

One of the spectacular consequences of the dense interlocking ice crystals is the blue color that we see in freshly broken icebergs and glaciers. Sunlight includes all of the colors, or wavelengths, of visible light. The low-energy wavelengths, such as red and yellow, can be absorbed during travel through glacial ice. These absorbed colors cannot be seen in the surface. Higher energy wavelengths, primarily blue, are not absorbed, but instead exit the ice surface to reach the eye of the beholder. Glaciers frequently appear white rather than blue because air bubbles, fractures, and pitted surfaces from exposure to air and rain scatter light.

The destructive power of floating ice can be awesome. Their tremendous inertia enables large icebergs to tear through the side of a steel-hulled ship the size of the *Titanic*. In Southeast Alaska, many boaters have perilous encounters with ice. My own experiences are somewhat less dramatic than the *Titanic*, but exciting nonetheless. During the summer of 1983, Kim Ouderkirk and I borrowed a boat to conduct geological research in the Tracy Arm area. During the first week, we took an eight-mile boat trip to map the geology along Tracy Arm. Before long the outboard motor quit and would not restart. Two hours of my mechanical tinkering proved useless and we were forced to consider other options. In 1983, few boats ventured up the fjord, so we somewhat reluctantly began to row. Unfortunately, a combination of inadequate length oars and poor oarsmanship quickly resulted in a broken oarlock and a mere three miles of progress. We noticed that the icebergs around us were floating toward the open water of Holkham Bay more rapidly than we were. The tide was pulling the bergs more quickly than it pulled our boat because bergs are at least 70 percent beneath the water. It occurred to me that we could simply rope our boat to the nearest berg and lie back to enjoy the scenery as the berg towed us toward Holkham Bay (Figure 6). Suddenly, the fast approach of a second, larger iceberg awoke us from our relaxation. The second berg was traveling far more rapidly than the berg to which we were attached because it was larger and had more surface area beneath the water. The two bergs were on a collision course with our boat between them! Luckily, we were able to untie the rope and clear the boat from danger. After that we chose to rely on tried-and-true modes of

FIGURE 6. Kim Ouderkirk explaining the use of icebergs for rapid transit by the tide in Tracy Arm.

transportation and fashioned a sail from the rain fly of our tent and one of the oars, safely arriving at Harbor Island two days later.

Glacial Flow and Associated Landforms

The movement, or flow, of glacial ice is analogous to the flow of water in rivers, but the rate of movement is much, much lower. Rivers generally flow at velocities of feet per second, while glaciers typically flow at velocities of a few feet to tens of feet per day. Glaciers leave behind evidence of their movements, from historical ice positions to the landscapes they have carved and eroded. Scientists have made major advances in our ability to read this intriguing history and unlock clues about the geological past (see "Measuring Glaciers," page 17).

Two types of movement occur in glaciers: basal slip and plastic flow. Basal slip is simply the sliding of ice along the rock below. This process is aided by the presence of water that may result from seepage of meltwater to the base of the ice or melting from frictional heating. The accumulation of a large amount of water along the base of a glacier reduces the friction between ice and the underlying rock and can cause a brief surge in the speed of glacial movement to several hundred yards a day.

Plastic flow is the movement or sliding of ice molecules past one another within the ice crystals. This type of movement can occur only when the pressure on ice crystals equals that produced by about 150 feet (46 m) or more of overlying ice. Because this process involves flow within solid material, it is considerably slower than the flow of water in a stream or basal slip. Plastic flow is fastest in the central areas of glaciers that are little slowed by friction with rocks beneath the glacier. In the case of alpine or valley glaciers, where friction created by valley walls slows the flow of ice along the sides of the valley, flow is fastest in the center of

Measuring Glaciers

Curious individuals and scientists have been measuring the velocity of glacial flow since the early eighteenth century. A standard method was to survey the changing positions of stakes driven into the surface of the ice. For example, students from Petersburg High School and the University of Alaska have measured the velocity of LeConte Glacier for several years, creating a long-term record of ice flow.

Today, more sophisticated measurements are collected with global positioning system (GPS) receivers that are placed on the ice and triangulate their position by receiving signals from satellites. Positions and times are recorded in order to determine velocities and directions of movement. Because this technique requires expensive, cutting-edge equipment, it is used only in a few parts of the world, mostly on high-velocity ice streams in Antarctica.

Although these velocity data are useful for understanding flow and help in understanding variation in ice volume with time, additional information is needed in order to make reasonable projections of ice volumes and sea level. Changes in ice volume over time dramatically impact sea level. Volume estimates require data on area and thickness of ice. Areas are relatively easy to calculate from the large number of air photos and satellite images that are available. However, thickness is much more difficult to measure because it requires data from ice-penetrating radar, high-resolution *seismographs*, and expensive drill holes. The radar and seismic techniques rely on detecting energy produced at the surface, reflected off the rock-ice interface underneath the glacier, and then returned to the surface. The time for the signal to return to the surface is proportional to the depth of ice. Recently, researchers have developed airborne-laser altimetry for precisely measuring the changes in elevation and shape of the ice surface. These data can be combined with areas of the glaciers in order to compute loss of ice. If we extrapolate them into the future, they can predict sea level changes.[10]

the channel. As a result, alpine glaciers flow fastest in the upper part of the central channel where there is little or no friction. The variation in velocity within the channel, combined with merging streams of ice and irregularities in channel shape, can cause complex patterns of flow that create folds in the ice layering (Figure 7). Because plastic flow can only occur under substantial pressure, ice in the uppermost part of glaciers is brittle and fractures down to the depth at which plastic flow is occurring. These fractures, called *crevasses*, can extend more than 100 feet (30 m) down into the glacier, presenting treacherous obstacles to travel.

FIGURE 7. Folds in ice at the terminus of Margerie Glacier, Glacier Bay. The dark lines are rock-rich layers probably formed from rock falls and streams that carry debris onto the ice. These layers are buried by snow and then fold as ice carries the rock downhill. Complex patterns result from variable velocity due to friction along the base and sides of the channel, flow over uneven bedrock, and slippage along fractures as the ice flows.

Geologists have long recognized the many types of erosion and deposits associated with glaciers. Study of the shapes or forms of the earth's surface and the processes of erosion and *deposition* that cause these formations is called *geomorphology*. Glacial erosion is caused by removal or scraping of rock or soil by moving ice and by the rock material that the ice carries. Erosional features range from large-scale landforms, such as *U-shaped valleys* and *fjords*, to small *glacial striations* or grooves. Glacial deposition occurs when the eroded material is dropped from the glacier as a result of irregularities in the flow or melting of the ice. By identifying glacial erosion and deposition we can determine the position and behavior of ice in the past.

Glaciers have cut distinctive shapes into the mountains and coastline of Southeast Alaska. High in the mountains, *cirques* form when the head of a valley glacier carves a steep-sided bowl-shaped depression into solid rock. When the ice melts, flat-bottomed bowls surrounded by cliffs are left behind. These picturesque features are generally located near the summit of mountains and may contain small lakes called *tarns*. Three or more glaciers flowing away from one another can carve steep valleys, leaving nearly vertical peaks called *horns* in the center. Perhaps the best-known horn in the world is the Matterhorn in Switzerland. Devil's Thumb, east of Petersburg, is a particularly impressive horn in Southeast Alaska.

U-shaped valleys are cut into mountains by alpine glaciers as they grind and carve their way down slope. The name is derived from the shape of the valley, which has nearly vertical sides and a broad, nearly horizontal floor. These U-shaped valleys are easily distinguished from valleys cut by running water, which instead have a V-shape with shallower, sloping sides and a narrow floor. Glacial valleys that are flooded by the ocean are called fjords. These snaking arms of the sea are often long, narrow, winding passageways (Figure 8) that may be more than a thousand feet

FIGURE 8. Tracy Arm fjord viewed from about 3,000 feet (915 m) above sea level on the south side of Sweetheart Ridge. The 6,666-foot (2,020-m) summit of Mt. Sumdum is in the clouds along the horizon.

FIGURE 9. Striation-covered valley wall scoured by rocks that were dragged across the surface by Sawyer glacier in Tracy Arm. The size of these striations attests to the tremendous forces imposed on the rock during glacial flow.

(300 m) deep and often end in a glacier. Tracy Arm, about 50 miles (80 km) south of Juneau, is more than 1,200 feet (360 m) deep. Knife-sharp ridges called *arêtes* often separate parallel glacial valleys. Sweetheart Ridge, separating Tracy Arm to the south from Sweetheart Lake to the north, is an example of an arête.

Linear grooves, or striations, gouged by glaciers dragging loose rocks, are common along the sides of fjords. In Tracy Arm (Figure 9), for example, these subhorizontal striations can be found tens or hundreds of feet above sea level (more than a thousand feet above the bottom of the deep fjord), indicating the great depth of the ice when the grooves were carved.

Glaciers can erode large quantities of rock and transport it over a huge distance. Rock that is transported by a glacier and then directly deposited by the ice is called *till*. Till has two distinctive characteristics that distinguish it from stream deposits. First, because ice can support and carry virtually any size of rock fragment, it contains rocks that are far more varied in size than material from streams. All of these rock fragments are carried and deposited together (unsorted by size). Thus, a variety of rocks deposited together, unsorted by size, is a likely indicator of glacial transport and deposition. In fact, glacial deposits may include huge boulders called *erratics* that may be tens of feet in size. A second characteristic of till is that it is generally not layered, unlike stream deposits. Till is deposited in sheets, ridges, or hills called *moraines*. These vary considerably in shape and size, depending on where and how they are deposited. Four of the most common types of moraines are terminal moraines, occurring at the farthest extent of ice flow; end moraines at the end of the glacier; lateral moraines along the sides of the glacier; and medial moraines within the ice stream. Exceptions to the characteristic lack of layering in glacial till occur when it is redeposited by meltwater after deposition by a glacier. This material, called "stratified drift," has some characteristics of till, but rock fragments tend to be somewhat sorted by size and deposited in layers. Large deposits of stratified drift are well exposed along the shorelines of Glacier Bay.

Ice Age History of Coastal Alaska

Southeast Alaska was probably almost completely covered by ice sheets at several different times during the Pleistocene Epoch. The most recent ice age, known as the Wisconsin Ice Age, ended about fourteen thousand years ago in Southeast Alaska. Until recently, most researchers assumed that all but the highest mountain peaks were encased in ice during this ice age. However, animal bones found in caves on Prince of Wales Island suggest that there was an ice-free swath of land along the coast of Southeast Alaska. The full extent and duration of this ice-free zone is still unknown. After the Wisconsin Ice Age many of the glaciers in Southeast Alaska may have advanced significantly during a cold period known as the Younger Dryas (12,750 to 11,750 years ago) and are known to have advanced during the Little Ice Age (about AD 1400 to 1750).

In Southeast Alaska, caves on Prince of Wales Island provided a safe haven for humans and animals during the ice ages. The island's large swaths of *limestone* bedrock are ideal areas for cavern formation because limestone is readily dissolved by rainwater that seeps into the rocks and erodes large cavities. Recent discoveries by archeologists indicate that humans used these caves at least ten thousand years ago.[11] In addition, older seal and bear remains in the caves suggest that at least some areas may have remained free of ice during at least part of the last ice age. The evidence for humans living west of the Coast Mountains ten thousand years ago is consistent with the theory that humans may have crossed the Bering Sea before this—when low sea levels created a "land bridge" from Siberia—and moved southward along a coastal ice-free corridor.

CHAPTER 3

Plate Tectonics: A Geologic Paradigm

As early as 1596, the Flemish geographer Abraham Ortelius recognized evidence for the movement of the earth's continents.[12] Subsequently, evidence was documented by Reverend Thomas Dick (1800s), Antonio Snider-Pellegrini (1858), and more recently by Frank Taylor (1910) and Alfred Wegener (1912). All of these observant individuals noted that the shorelines of continents, particularly Africa and South America, could be fit together like a jigsaw puzzle. This led them to hypothesize that these two continents, now far apart, had at one time been joined. Initially considered absurd by many scientists, the idea that continents are constantly moving opened the way for the development of the theory of plate tectonics in the 1960s. This theory could not have developed were it not for the wealth of data on the topography and *geophysics* of the sea floor collected during and following World War II. The sea floor topography and magnetic signatures in the rocks showed linear ridges, known as midocean ridges, in the midst of ocean basins. *Minerals* in the rock preserve the polarity (direction to the north magnetic pole) of the earth's magnetic field that has changed with time. The midocean ridges are boundaries between plates of oceanic crust that are symmetrically striped with parallel bands of alternating polarization.

Geologists first hypothesized and then confirmed that new volcanic rock emerges at midocean ridges, forming new oceanic crust. In other words, the midocean ridges are *plate boundaries* where crustal plates are forming and moving apart.

After recognizing the formation of new oceanic crust at midocean ridges, researchers began extensive studies of magnetism in continental rocks. Many rocks preserve the orientation of a compass needle (see "Earth Magnetism," next page). Therefore, the latitude of the rock during formation can be obtained if the original orientation of the rock is known and the magnetic signature has not been altered. The *paleomagnetic* signatures and the ancient fauna imply that some rocks in the cordillera originated thousands of miles south of their current location.

Today, plate tectonic theory is so widely accepted by scientists that it is the fundamental theoretical framework for geology. Geologists describe the earth's surface in terms of six large crustal plates and several smaller ones that are moving relative to each other and relative to any fixed reference point on the earth. These plates move at rates of as much as 4 inches (10 cm) per year. Rates of movement can be measured with modern surveying techniques and by interpretation of seafloor magnetic data.

The two types of crustal plates are continental and oceanic. Continental plates are 15 to 60 miles (25 to 100 km) thick and composed of rocks that are rich in silicon, aluminum, and potassium, and poor in iron and magnesium. Oceanic plates are 3 to 5 miles (3 to 8 km) thick and composed of rocks rich in iron and magnesium, and poor in silicon, aluminum, and potassium. Plates are separated from each other by three types of boundaries:

Divergent boundaries occur where plates move away from each other and new crust is added through volcanic eruptions that result from partial melting of rocks beneath the earth's crust, in the *mantle* (Figure 10). These boundaries form long linear ridges,

Earth Magnetism: A Tool for Determining the Ancient Position of the Crust

Igneous and sedimentary rocks can retain a record of the earth's magnetic field orientation and polarity when they form (paleomagnetic signature), which can help to determine the latitude of formation. This occurs because the earth has a dipolar magnetic field with force lines that are essentially vertical at the north and south magnetic poles and horizontal at the equator (see diagram below). The magnetic field has switched polarity many times during the earth's history. During times of reversed field a compass would point to the southern end of the planet. However, the force lines shown below would still have the same orientation. When *minerals* in igneous rocks cool to below what is called the Curie temperature, they preserve the orientation of the magnetic field. Therefore, igneous rocks formed near the equator can retain a subhorizontal magnetic signature. Paleomagnetic signatures and patterns of seafloor magnetism have been used to construct a history of plate movements and positions for the Paleozoic and Cenozoic eras of earth history. Many igneous rocks in Southeast Alaska contain evidence for subhorizontal magnetic fields when they cooled. This indicates an origin near the equator or subsequent rotation.

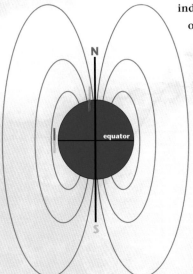

SCHEMATIC DIAGRAM of the earth's magnetic field and the orientation of compass needles or magnetic alignment of mineral grain in rocks at the equator (parallel to the earth's surface: yellow/orange line) and North Pole (perpendicular to the earth's surface: yellow/orange line).

or oceanic spreading centers, within ocean basins. For example, the North American and Eurasian plates are moving apart at the mid-Atlantic ridge. Volcanic rock along oceanic spreading centers is typically *basaltic* in composition (rich in iron and magnesium and poor in silicon). Earthquakes are frequent along these boundaries, but they tend to be shallow and low in magnitude.

Convergent boundaries occur where plates move toward each other (Figure 10). Convergent boundaries are subdivided into two types. In the first type, a subduction zone, a plate is consumed by sinking under the adjacent plate. The subducted oceanic plate sinking into the earth's hot interior triggers melting and volcanic activity on the overlying plate. The sinking plate is always oceanic, because oceanic crust is made of denser material than continental crust, whereas the overlying plate may be oceanic or continental.

The process of convergence tends to create a linear arc of volcanic islands when the overlying plate is oceanic, and a linear belt of volcanoes on a continental margin when the overlying plate is continental. An example is the collision of the Juan de Fuca oceanic plate and the North American continental plate along the West Coast, which created the Cascade volcanoes in California, Oregon, and Washington. Volcanic rocks above the subducting oceanic crust are typically *andesitic* (intermediate in iron, magnesium, and silicon content).

In the second type of convergent plate boundary, a collision zone, the complete consumption of an oceanic plate by subduction

► **FIGURE 10.** Diagram showing the relative movement of crustal plates at plate boundaries. Dark blue is oceanic crust, green is continental crust, and yellow and orange are volcanic rocks caused by melting associated with upwelling of mantle rock beneath divergent plate boundaries or subduction at convergent boundaries. Alternating shades of blue stripes in the oceanic crust adjacent to the midocean ridge in the divergent boundary indicate the alternating polarity of the paleomagnetic signature of the oceanic crust.

Divergent

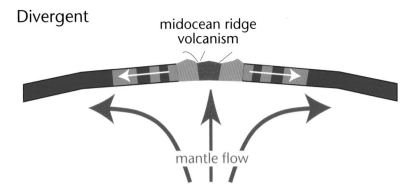

midocean ridge
volcanism

mantle flow

Convergent A: subduction of oceanic crust

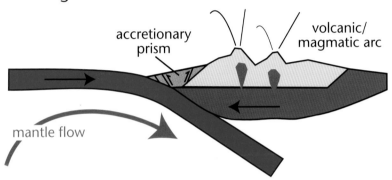

accretionary
prism

volcanic/
magmatic arc

mantle flow

Convergent B: continental collision

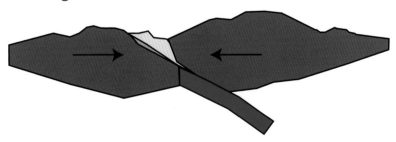

Transform

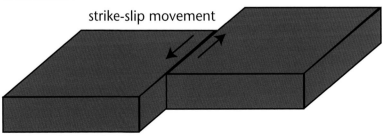

strike-slip movement

results in the collision of two continents or of a continent and a volcanic island arc. Neither of these is dense enough to sink into the earth's interior, so the crust at the boundary thickens, often building massive mountain ranges. An example is the Himalayan Mountains, where the Indian subcontinent is colliding with Asia. Lower crustal *metamorphic rocks*, formed deep in the earth, are brought to the surface or exhumed (by a process known as *exhumation*), *igneous rocks* are formed, and numerous large-magnitude earthquakes occur at these sites of collision.

Transform boundaries occur where plates slide horizontally past each other (Figure 10). The two plates may be continental or oceanic. In western California and farther north along the British Columbia coast, the Pacific plate is sliding northward past North America along the San Andreas and Queen Charlotte faults. Volcanic activity rarely occurs along these boundaries, though earthquake activity may be frequent.

The most widely accepted theory for movement of crustal plates combines two processes. First, the difference in temperature between earth's core and crust causes the mantle to move in a circulatory motion called convection. The earth's interior is hot because of both the initial heating during the formation of the planet and subsequent *radioactive decay*. Heat escaping from the interior of the earth drives convection cells that bring hot mantle material upward at the divergent plate boundaries. Some of the rising material melts and reaches the surface to make new oceanic crust. Second, the difference between the highly dense, old oceanic crust and the relatively less dense, younger oceanic or continental crust causes the dense older oceanic crust to sink. The combination of upwelling hot mantle at divergent boundaries and sinking cold oceanic crust at convergent boundaries moves the crustal plates across the earth's surface.

Oceanic crust is constantly forming at divergent plate boundaries. But consumption at convergent plate boundaries approximately balances this production, resulting in little change in the area of oceanic crust with time. Continental crust forms when the oceanic crust and sediments partially melt above subduction zones. This concentrates silica and other elements in the overlying crustal plate. Continental crust formed in this fashion is too low in density to sink into the mantle, so it simply moves about the earth's surface until it collides with another piece of continental crust. The plate tectonic process and evolution of the crust is analogous to a distillation process that makes continental crust instead of alcohol. As a result of this distillation, the amount of continental crust increases slowly with time.

Convection of the mantle is similar to the movement of soup when it boils on a stove. Expansion of the heated soup at the bottom of the pan causes it to decrease in density and rise to the surface. There, it cools from contact with the air. Cooling increases its density, sending it back toward the bottom of the pan. Vegetables in the soup are less dense and cannot sink, so they move about on the surface. Heat from the burner beneath the soup pan is analogous to heat in the earth's interior. This heat escaping from the earth's interior is caused by the decay of radioactive elements and possibly remnant heat from planet formation. Although the mantle is composed of solid silicate rock, it behaves like a very viscous fluid, flowing slowly over time. The rocks partially melt as they expand and flow upward toward the surface. The resulting basaltic *magma* flows more rapidly than the solid rock and reaches the surface at divergent plate boundaries, or oceanic spreading centers, where it forms oceanic crust. The oceanic crust moves away from the divergent boundary and cools, eventually becoming dense enough to sink back into the mantle at a convergent plate boundary. Sinking or subduction of oceanic crust results in partial melting that tends to produce rocks with a different chemical composition and lower density than the basalt. These "continental" rocks are lower density than the mantle and behave a bit like vegetables in the soup; they cannot sink. Thus, the continents are condemned to move about at the whim of the active mantle below.

Geologic Time and Tectonics

The plate tectonic paradigm explains the movement of large pieces of crustal real estate across thousands of miles. Recent measurements, partly based on GPS data, indicate that parts of the earth's crust move at absolute velocities of 5 inches (12 cm) per year or more.[13] Thus, we can estimate that the American and African plates moved on the order of 1,500 miles (2,400 km) since rifting, and at this velocity it would have taken the Atlantic approximately 200 million years to reach its current width. While imprecise, these estimates suggest that the crustal plates have moved impressive distances.

The ages of geologic events are derived from the radioactive decay of unstable isotopes of elements such as uranium. Isotopic ages of meteorites that were probably formed at the same time as the earth and the oldest earth rocks suggest that the earth is about 4.6 billion years old. Many different types of geologic events can be dated with isotopes: for example, the crystallization of molten rock or magma, thermal events in the crust or *metamorphism*, and cooling events during uplift and erosion. Examples of isotopic dating by uranium decay to lead and samarium decay to neodymium are described below.

Long before the widespread use of isotopic dating began, geologists constructed a relative time scale for the planet with numerous

► **FIGURE 11.** Geologic time scale. Modified from the Geological Society of America Time Scale (1998), and Monroe and Wicander (2001).

Eon	Era	Period	Epoch	Major events	Millions of years ago
Phanerozoic	Cenozoic	Quaternary	Holocene/ recent	post–ice age retreat of glaciers	0.01
			Pleistocene	ice ages; most recent volcanism from Mt. Edgecumbe	1.8
		Tertiary	Pliocene	first humans	5.3
			Miocene	youngest plutons in Southeast Alaska	23.8
			Oligocene		33.7
			Eocene	formation of Himalayan and Alpine mountains	54.8
			Paleocene	continued orogeny in Southeast Alaska	65
	Mesozoic	Cretaceous		major episode of mountain building in the Coast Mountains, Southeast Alaska; extinction of dinosaurs	144
		Jurassic			206
		Triassic		first mammals and dinosaurs	248
	Paleozoic	Permian		deposition of limestone in the Alexander ter-rane now exposed on Kuiu Island; last moun-tain building in the Appalachians	290
		Pennsylvanian		coal-forming swamps	323
		Mississippian		first reptiles	354
		Devonian		first amphibians	417
		Silurian			443
		Ordovician		first land plants	490
		Cambrian		first fish	540
Precambrian				oldest rocks in the Coast Mountains: pebbles in conglomerate	2500
				first fossils	4000
					4600

subdivisions that are largely based on the nature of fossil evidence for life found in *sedimentary* rocks of each division. Divisions have subsequently been dated with isotopes to provide absolute ages. For example, significant tectonic activity in the Coast Mountains of Southeast Alaska and British Columbia took place near the end of the Mesozoic (age of reptiles) and beginning of the Cenozoic (age of mammals). Isotopic data indicate that this transition took place about sixty-five million years before present. A simplified version of the geologic time scale is presented in Figure 11.

CHAPTER 4

Rock Types:
Classification of Earth Materials

M aterials are often classified according to their melting temperature. For example, rocks melt at high temperatures (over 1,292°F or 700°C), ices melt at intermediate temperatures (near 32°F or 0°C), and gases melt at low temperatures (much less than 32°F). Rocks are composed of numerous grains of one or more minerals. These minerals are crystalline solids with a clearly defined range of chemical composition. Many minerals are familiar to all of us (for example, halite or salt, *quartz*, and gold), but others are less well known (for example, *feldspar*, *garnet*, and *olivine*). There are over three thousand known minerals, but only a relatively small number of them are abundant. In fact, quartz and two types of feldspar make up more than 60 percent of the earth's continental crust.

Rocks are described in terms of their mineralogy and texture. These observations combined with direct observation of surface processes (for example, deposition of sediments and eruption of volcanoes) are used to interpret their origin as igneous, sedimentary, or metamorphic.

Igneous rocks are formed when molten or liquid rock (magma) cools and solidifies. Hence they are named for the Latin word "ignis," meaning fire. Magma may solidify or crystallize below the earth's surface and form *plutonic* (or intrusive) rocks, or it may erupt (reach the surface) and form volcanic (or extrusive) rocks. Igneous rocks cover a small amount of the continental surface, but are prevalent at the ocean floor. They are classified by their texture, which depends mostly on the way in which they cooled and their chemistry. Volcanic rocks cool rapidly. As a result, they are generally so fine-grained that their crystals are too small to be seen without a microscope. The coarser-grained plutonic rocks result from slower cooling, and have crystals large enough to be seen without magnification. Igneous rocks are also classified by their chemical composition, which depends largely on the material that is melted (for example, oceanic or continental crust).

Silicon dioxide (silica) is the most important chemical component used to classify rocks. For example, *mafic rocks*, such as basalt, contain low concentrations of silica, while *felsic rocks*, such as *rhyolite,* contain high amounts of silica. Coarse-grained intrusive rocks with *intermediate* silica content, such as *diorite* (Figure 12), and fine-grained extrusive rocks with intermediate silica content, such as andesite and andesitic basalt, are the most common igneous rocks in Southeast Alaska, because these rock types commonly form in subduction zones.

Sedimentary rocks are formed from weathered material or sediment at the earth's surface. Hence, they are named for the Latin word "sedimentum," meaning settling. Sedimentary rocks may form from settling of fragments of weathered material or precipitation of minerals from water. *Clastic* sedimentary rocks (Figure 13) form by solidification of fragments from pre-existing rocks (for example, compaction and cementation of sand forms sandstone). Clastic sediments and the resulting rocks may be immature—little altered from the original rocks (for example, *greywacke*)—or mature—chemically or mechanically altered from the original rocks (for example, quartz sandstone).

FIGURE 12. Coarse-grained diorite in a body of intrusive igneous rock called a pluton, George Island (location shown in Figure 1). The angular dark green mineral is *hornblende* and the white mineral is *plagioclase*. Two veins of red garnet crystals cut across the diorite.

Chemical sedimentary rocks form when water from streams, lakes, or oceans precipitates dissolved ions. They may be inorganic, as when evaporating water leaves behind rock salt, or organic, as when animal remains are solidified to make fossiliferous limestone. Clues about the origin of sedimentary rocks can be found by comparing features in the rock to those found in modern environments. For example, in modern settings, certain groups of sediments, including carbonate reef deposits, only form in shallow warm oceans. Thus, ancient limestone containing fossil organisms like those in reefs is inferred to have formed similarly. Although sedimentary rocks make up a small proportion of the earth's volume, they cover much of the surface and contain numerous economically important materials like coal and hydrocarbons. Many types of clastic sedimentary rocks are common in Southeast Alaska. These rocks tend to be immature because many of the sediments have been through only one cycle of erosion and transport, and therefore retain many of the minerals common to the parent rock from which they were derived.

Metamorphic rocks are formed by solid-state changes in the mineral composition or texture of pre-existing igneous, sedimentary, or metamorphic rocks. In other words, they formed not from melting, but as a result of high temperature and pressure. Hence they are named using the Latin words "meta" (change) and "morphe" (form). These rocks typically form deep in the earth because temperatures sufficient to cause mineralogical changes are almost entirely restricted to the mid-crust and greater depths.

FIGURE 13. Fine-grained clastic rock known as mudstone from southern Prince of Wales Island (location shown in Figure 1). The thin light and dark green layers are very small fragments of rock deposited by flowing water. Wavy layers and truncations indicate that the current caused ripples along the bottom.

Metamorphic rocks are classified by their chemistry and texture. The most important textural classification is the presence or absence of layering, or *foliation,* caused by deformation and recrystallization. Foliated rocks include *slate* (fine-grained with a closely spaced foliation), *schist* (coarse-grained with a closely spaced foliation), or *gneiss* (coarse-grained with a coarsely spaced foliation) (Figure 14). Although metamorphic rocks must make up much of the earth, their presence at the surface is largely restricted to

FIGURE 14. Folded layers in gneiss from the Coast Mountains south of Juneau (Coast Plutonic Complex), Southeast Alaska. Several features that crosscut each other allow reconstruction of the rock's history. The gneissic layering (light brown and tan) formed first from the alignment of minerals during metamorphism. Subsequently, forces imposed on the rock folded the gneissic layering. These forces eventually broke the rock along a subhorizontal plane, or fault, which displaces layering in the upper half of the specimen. After faulting, the rock broke again along a plane from lower left to upper right. This fracture filled with magma, which crystallized to make the speckled band of igneous rock called a dike.

mountain ranges and some of the older parts of continents (for example, the Canadian Shield).

Metamorphic rocks are derived from igneous and sedimentary rocks by heating during burial (regional metamorphism) or heating from nearby igneous rocks (contact metamorphism). Both of these types of metamorphic rocks are common in Southeast Alaska. They are the most abundant rock type for much of the Coast Mountains and some of the islands.

Low-lying areas of the earth's surface, particularly lakes and oceans, are dominated by sedimentation and thus by sedimentary rocks. However, mountainous areas of the earth, such as Southeast Alaska, form as a result of voluminous igneous activity or uplift and exhumation of deep, high-temperature rocks. Therefore, mountains tend to have central areas underlain by igneous and metamorphic rocks that have been uplifted relative to the surrounding areas underlain by sedimentary rocks.

CHAPTER 5

Faults, Earthquakes, and Mountain Building

When force is applied to rocks, they are able to absorb some energy because they are somewhat elastic, and thus deform only slightly. Application of a very large force, however, will cause the rock to break along a fault. This is analogous to stretching an elastic or rubber band. A small force stretches it and a greater force will cause the band to snap or fail. Faults, then, are fractures in rock with displacement across them.

When rocks break along a fault, energy is released in the form of an earthquake. Studying the transmission of this energy through the earth is important for understanding earthquakes. Likewise, earthquake studies provide important information about the nature of the earth's interior. The crust beneath Southeast Alaska frequently quakes with the energy released along faults. The principal culprit is the Queen Charlotte–Fairweather fault located offshore along the west coast of Baranof and Chichagof islands and intersecting the mainland west of Glacier Bay (Figure 17). Most of the other large-scale faults in Southeast Alaska are historically inactive, including the Chatham Strait fault, which separates Baranof and Chichagof islands from Admiralty Island, and faults in the Coast shear zone, which parallels the western edge of the Coast Mountains.

The Floating Crust: Isostacy

The distribution of ocean basins, continental plains, and mountains can be attributed to different types and thicknesses of crust floating on the underlying mantle. Seventy-one percent of the earth is covered by water. If this volume of water were equally distributed over the surface, it would be about 1.5 miles (2.5 km) deep. However, because rocks beneath the ocean basins are far lower in elevation than those of the continents, the oceans average 2.25 miles (3.5 km) in depth and the continents range from slightly below sea level (adjacent to shorelines, continental shelves, and in isolated inland areas) to more than 29,000 feet (8,700 m) above sea level (Mt. Everest). This variation in height results from crustal and upper mantle rocks floating on lower mantle rocks that flow slowly like Silly Putty. Crustal rocks are, on the average, lower in density than the mantle below and therefore float. However, crust has variable density: the continental crust has an average density of about 172 pounds per cubic foot (2.75 g/cm^3) and the oceanic crust has an average density of about 185 pounds per cubic foot (2.95 g/cm^3). This variation in density is one reason for oceanic crust lying below sea level and continental crust lying at higher elevations. However, there is another reason for oceans and continents—geophysical data tell us that the oceanic crust is not only higher in density than continental crust, but also approximately one-fifth the thickness.

The crust floating in the mantle is analogous to logs floating in water (Figure 15). Dry logs that contain very little water have a

► **FIGURE 15.** Diagram comparing logs floating on water and the earth's crust floating on the mantle. The solid earth's crust (analogous to a log) floats on the viscous, more dense mantle (analogous to water). Variable crustal thickness and variable density (shown by different colors) affects the resulting height of the crust above the mantle and sea level, producing ocean basins, low plains, or mountains.

Logs floating on water

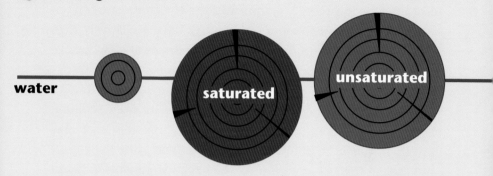

water   saturated   unsaturated

Crust with variable thickness "floating" on the mantle

mantle

Crust with variable density "floating" on the mantle

mantle   high density   low density   lowest density

relatively low density and float high in the water. These logs behave much like the low-density granitic rocks of the continents. Water-saturated logs that contain a lot of water have a relatively high density and float low in the water. These high-density logs behave much like the higher density basaltic rocks beneath the oceans.

This is a reasonable explanation for the oceans and continents. However, is this the only explanation, and why does elevation vary within the continents? The answer is that the crust may vary in thickness in addition to density. Areas with thick crust tend to float higher—for example, the Rocky Mountains in North America—and areas with thin crust float lower—for example, the central plains east of the Rockies.

The analogy with logs floating in water can be taken a step further. All other factors taken as equal, the top of a thick log will protrude higher above the water than the top of a twig. The thick crust beneath the continents is like the log and the thinner crust beneath the oceans is like a twig. So, two explanations exist for the high elevation of mountains: the continental crust is both less dense and thicker than the oceanic crust.

Isostacy is a theory that deals with floating of the earth's *lithosphere* on rocks below in a hypothetical state called "isostatic equilibrium." Because the earth is dynamic, ice or rock can be added or removed, causing the crust to move upward or downward with respect to sea level. Areas of the continents that have recently been glaciated rise upward with respect to sea level in a process called "isostatic rebound." This can be inferred from raised shorelines along coasts and can be measured by accurate surveying techniques. Uplift rates for recently glaciated mountains in the Glacier Bay area of Southeast Alaska are presented in Figure 21.

Mountain Building

The theory of isostatic equilibrium provides an explanation for high mountains, low plains, and ocean basins. But it does not explain why the crust differs in density or thickness from one location to another. Oceanic and continental crust differs in composition and density as a direct result of plate tectonics. But why are parts of continents mountainous and other parts low-lying plains? Isostatic equilibrium requires that high mountains be underlain by low-density crust and/or thicker crust. Low-density crust may explain the height of some mountains; however, it is likely that most high mountains are underlain by crust which is thicker than adjacent areas. Crust may be thickened by addition of magmas, and folding and faulting: tectonic thickening. All of these processes are important along convergent plate boundaries (Figure 10), where magma is produced during subduction and the crust is shortened and thickened as two plates collide. Addition of magma can be observed directly where lavas erupt at the surface and form volcanic mountains like those of the High Cascades. However, *intrusion* of magma to form plutons may also thicken the crust. Both magma eruptions and intrusions and tectonic thickening played important roles in the growth of the Coast Mountains in Southeast Alaska.

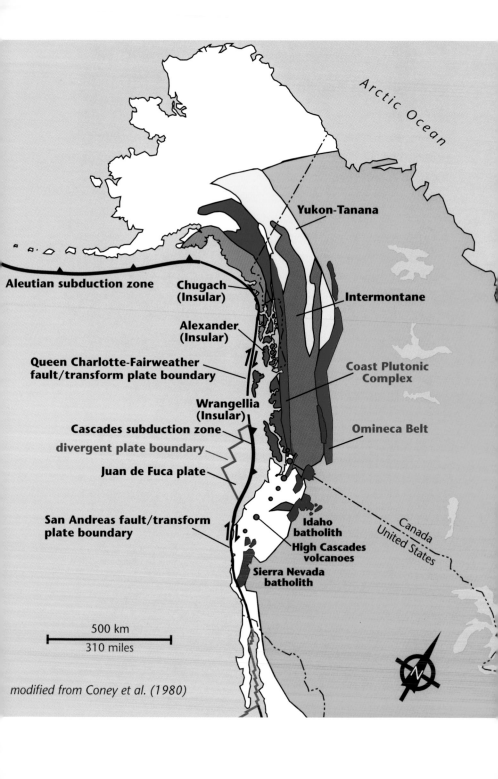

Arctic Ocean

Yukon-Tanana

Aleutian subduction zone

Chugach
(Insular)

Intermontane

Alexander
(Insular)

Queen Charlotte-Fairweather
fault/transform plate boundary

Coast Plutonic
Complex

Wrangellia
(Insular)

Cascades subduction zone

divergent plate boundary

Omineca Belt

Juan de Fuca plate

San Andreas fault/transform
plate boundary

Idaho
batholith

High Cascades
volcanoes

Canada
United States

Sierra Nevada
batholith

500 km
310 miles

modified from Coney et al. (1980)

CHAPTER 6

The Alexander Archipelago

S hortly after the development of plate tectonic theory, scientists recognized that the west coast of North America is a complex plate boundary. The boundary is transform from Baja California to northern California and from southern British Columbia to Southcentral Alaska, and convergent from northern California to southern British Columbia (Figure 16).[14] The identification of fault-bounded blocks of crust, known as exotic terranes, in the northern *cordillera* of British Columbia and Southeast Alaska led to a revolution in thought during the 1970s. These widespread terranes have no obvious relationship to the adjacent rocks and locally contain evidence for deposition far from their current location.

In fact, much of southern and Southeast Alaska is made up of rocks that are thought to have been transported great distances northward to their present locations. Evidence for movement of

◄ **FIGURE 16.** Simplified tectonic map of western North America from Baja California to Alaska showing major geologic terranes and boundaries in Oregon, Washington, British Columbia, and Southeast Alaska; oceanic plates; and major plate-bounding faults. Important plutonic and metamorphic belts, including the Coast Plutonic Complex and Omineca Crystalline Belt, are shown in red.

the terranes includes the paleomagnetic signatures of the rocks and
fossils that are not found in adjacent rocks of North America. The
presence of subtropical and tropical fossils led geologists to pos-
tulate that rocks in Alaska and parts of adjacent western Canada
were deposited near the equator. Later research revealed that the
magnetic fields of many of these rocks match the earth's magnetic
orientation at the equator. One explanation is that the rocks traveled thousands of miles north when large blocks of crust moved
along transform faults. Another interpretation is that large blocks
of crust were rotated, disturbing the magnetic orientations. Largely
as a result of corroborating evidence, it seems likely that the first
interpretation is correct and at least some of the rocks have been
displaced great distances northward. The Wrangellia terrane is one
of the best-defined and delineated terranes in the cordillera. Its
rocks, currently found in widely disparate areas, including southeastern Washington, southern British Columbia, Southeast Alaska,
and the Wrangell Mountains of Southcentral Alaska (Figure 16),
are believed to have formed near the equator about 200 million
years ago.

The northern cordillera of British Columbia and Alaska comprises two high mountain ranges (the Coast Mountains and Rocky
Mountains), the core of which is underlain by metamorphic and
plutonic rocks of the Coast Plutonic Complex and Omineca Belt
(Figure 16). Belts of less metamorphosed rocks lie west of the
Coast Mountains, in the Insular superterrane, and between the
high mountains in the Intermontane superterrane. The Insular
and Intermontane superterranes are characterized by a number
of continental fragments, oceanic crust, and island-arc complexes
of various ages. Paleontological, geophysical, structural, and rock
composition data have been used to infer that many of these terranes are unrelated to rocks of North America and may have originated thousands of miles south of their present position.

The Coast Plutonic Complex and Omineca Belt include exhumed parts of the mid- to lower-continental crust that was thickened by the collision of terranes (Insular and Intermontane superterranes) with each other or North America during the last 200 million years. Metamorphic and igneous rocks, which formed during collision, from both superterranes and North America are now exposed in these two long and spectacular mountain belts.

Many rocks preserve the orientation and polarity of the earth's magnetic field during crystallization or deposition (magnetic signature), which is similar to preserving the orientation of a compass needle (see Earth Magnetism, page 25). The latitude of the rock when the magnetic field was preserved can be obtained if the original orientation of the rock is known and the magnetic signature has not been disturbed. Scientists have thus used the paleomagnetic signature of rocks and their ancient fauna to infer that some of the rocks in the cordillera originated thousands of miles south of their current locations. Comparison of these data wth the paleomagnetic signature of the rocks to the east, in the central part of North America, led to the landmark interpretation that many rocks in the cordillera may have formed far south relative to the rest of North America. Therefore, the northern cordillera is a collage of exotic terranes with differing ages and geographic origins (Figure 16).

The timing of fault movement, sediment deposition, and magma intrusion are critical for identifying terranes and deciphering the history of their movement and *accretion*. In addition, geochemistry, particularly isotope data, can be used to identify terranes. For example, trace concentrations of elements in volcanic rocks may provide clues about the rocks' origin and about relationships between rocks.[15] Isotope ratios can help to determine the timing of important events such as crystallization of minerals that retain paleomagnetic signatures and accretion of terranes to North America.

The Insular Superterrane: Assembly of a Composite Crustal Fragment

The westernmost part of the northern cordillera is underlain by a group of terranes that were assembled into what is called the Insular superterrane before accretion to North America. This superterrane includes some of the best-studied and understood terranes in the cordillera. Three of these terranes are parts of volcanic island arcs: the Alexander and Wrangellia terranes and the Gravina belt. The Alexander and Wrangellia terranes formed south of their current locations and moved northward along faults before colliding with North America between 165 and 100 million years ago. The Gravina belt is one of two younger terranes within the Insular superterrane, and it formed in a marine basin along the eastern edge of the superterrane. The second of the younger terranes is the Chugach accretionary prism along the western side of the superterrane. This accretionary prism is a strongly deformed wedge of sedimentary and metamorphic rocks that formed above oceanic crust that was subducting under the Insular superterrane during the late Cretaceous.

The Alexander and Wrangellia terranes are the two oldest volcanic island arcs in the Insular superterrane (Figure 17). The older Alexander terrane includes rocks that may be more than 500 million years old and rocks as young as 240 million years old. The younger Wrangellia terrane contains rocks that are 150 to 250 million years old. Most of the contact between these two terranes is faulted, obscuring the relationship between them. However, it appears that the younger Wrangellia terrane was deposited on

► FIGURE 17. Terrane map of Southeast Alaska, showing the Chugach, Wrangellia, and Alexander terranes, the Intermontane superterrane, subdivisions of the Coast Plutonic Complex, and the Coast shear zone. A = Admiralty Island, B = Baranof Island, C = Chichagof Island, K = Kupreanof Island, POW = Prince of Wales Island, and QC = Queen Charlotte Islands.

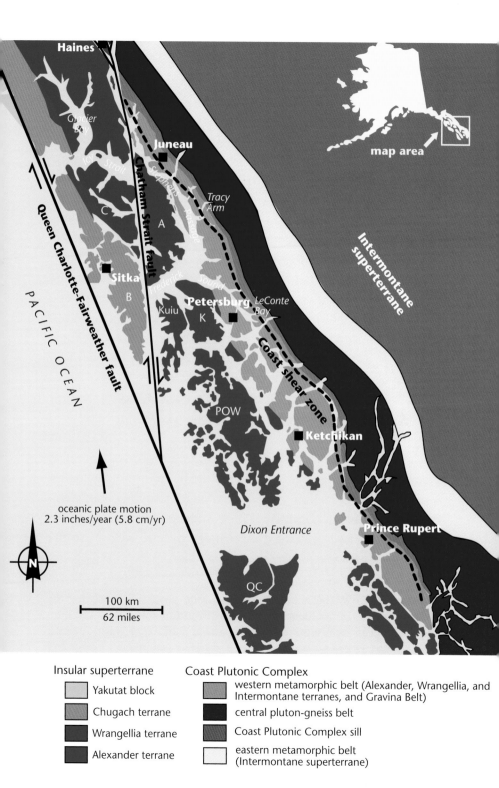

map area

Haines

Glacier Bay

Juneau

Tracy Arm

Sitka

Petersburg

LeConte Bay

Ketchikan

Prince Rupert

Dixon Entrance

Chatham Strait fault

Queen Charlotte-Fairweather fault

Coast shear zone

PACIFIC OCEAN

Intermontane superterrane

oceanic plate motion
2.3 inches/year (5.8 cm/yr)

N

100 km
62 miles

Insular superterrane

Yakutat block

Chugach terrane

Wrangellia terrane

Alexander terrane

Coast Plutonic Complex

western metamorphic belt (Alexander, Wrangellia, and Intermontane terranes, and Gravina Belt)

central pluton-gneiss belt

Coast Plutonic Complex sill

eastern metamorphic belt (Intermontane superterrane)

top of the Alexander terrane. Therefore, these terranes have likely been together for the last 250 million years.

Alexander terrane rocks are exposed on many of the major islands of the Alexander Archipelago (Figure 17) along the Alaskan panhandle. This terrane has a wide variety of sedimentary, volcanic, and metamorphic rocks that range in age from Ordovician to Triassic. These rocks have been intruded by early Paleozoic to Cenozoic granitic plutons on Prince of Wales Island and elsewhere. Early Paleozoic strata, well exposed on western Prince of Wales Island, include marine shale and chert mixed with layers of volcanic rocks.[16] Permian and Triassic shale and limestone and volcanic rock overlie the Paleozoic rocks on Kuiu, Kupreanof, and southern Admiralty islands. Early Paleozoic *granitic* rocks suggest that the Alexander terrane began as continental crust that rifted away from another land mass and then became the basement for subduction-related volcanism during the Mesozoic.

The oldest Wrangellia terrane rocks are late Paleozoic oceanic material (basalt, limestone, and *mudstone*) that formed about 15 degrees from the equator. These Paleozoic rocks are overlain by Triassic basalt, now largely metamorphosed to a rock known as *greenstone*, which can be found in British Columbia, Southeast Alaska, and Southcentral Alaska. The long history of oceanic volcanism and lack of continental rocks suggest that the Wrangellia terrane may have originated as a thick volcanic section on the ocean floor (oceanic plateau) or an island arc.

► **FIGURE 18.** Cross sections illustrating a vastly simplified model for the tectonic history of Southeast Alaska and British Columbia from 120 until 58 million years before present. The question marks after North America indicate that we cannot be certain whether the Intermontane superterrane was part of the North American continent at that time. The northwest margin of North America has been dominated by west-side-north fault motion (strike slip) after about 58 million years before present (see Figure 16). Left is west—Pacific Ocean—and right is east—North America.

West   **East**

120 mya: Gravina belt volcanism

a.

North America?

?   ?

120 to 101 mya: collapse of Gravina basin, collision of super-terranes, and subduction

b.

North America?

101 to 75 mya: accretion of Chugach terrane, thrusting, and emplacement of plutons and metamorphism

c.

North America?

72 to 58 mya: emplacement of Coast Plutonic Complex sill plutons

d.

North America

mya = millions of years ago

 fault with displacement

 Chugach terrane   Insular superterrane

Gravina belt   Intermontane superterrane

Coast Plutonic Complex   oceanic crust

Understanding the nature and orientation of these two volcanic arcs is critical for reconstructing the tectonic history and paleogeography of Southeast Alaska. Modern subduction zones, such as where the Juan de Fuca plate subducts under the High Cascades volcanic arc (Figure 16) accumulate an accretionary prism of sediment above the subducting oceanic plate (Figure 17). This is a bit like a bulldozer, with the overlying plate scraping sediment off the oceanic crust as the latter descends. The Chugach terrane (Figures 17 and 18c) was this type of prism during subduction of oceanic crust beneath a west-facing volcanic arc about 100 million years ago. The subduction polarity of the older volcanic arcs in the cordillera (the Alexander and Wrangellia terranes) is far more difficult to interpret, because the subduction ceased long ago and numerous later events have obscured much of the evidence that might be used to reconstruct the tectonic history. However, there are enough pieces of the puzzle to infer that one or more marine basins once separated terranes within the Insular superterrane.

The Gravina belt rocks, which are younger than Alexander and Wrangellia terrane rocks, once formed a narrow marine basin on the eastern side of the Insular superterrane (Figure 18). Clastic sedimentary rocks and volcanic rocks were deposited in this basin for about fifty million years, during the late Mesozoic. This basin closed between 90 and 100 million years ago, and arc *magmatism* ceased during contraction, which accompanied *folding* and faulting. The deformation that closed the basin likely caused some of the initial uplift of the Coast Mountains between about 100 and 50 million years before present (Figure 18).

Initial closure of the Gravina basin and magmatism in the Coast Mountains happened at the same time as the subduction of the Pacific plate beneath North America. As the basin closed, rocks and marine sediments were scraped off the subducting plate to form the Chugach terrane—a mixture or *mélange* of sedimentary, igneous, and metamorphic rocks.

CHAPTER 7

The Coast Mountains: Exhumed Lower Crust

I gneous and metamorphic rocks of the Coast Plutonic Complex (Figure 17) underlie the snow-capped peaks of the Coast Mountains. These rocks formed as a result of collision between the Intermontane superterrane of central British Columbia and the Insular superterrane of Southeast Alaska, and subsequent igneous activity during subduction of oceanic crust to the west. Metamorphic rocks in the complex include former sediments that were deposited at the earth's surface, buried to depths of up to about 20 miles (32 km), metamorphosed, and later exhumed by uplift and erosion.

Geologists have long known that deep crustal rocks are exposed in the Coast Mountains. However, only in the 1990s has detailed geophysical data on the thickness of the underlying crust been available. These data were obtained during a groundbreaking project, known as ACCRETE,[17] that integrated surface geology with deep crustal geophysical data. The variation of seismic energy velocities with depth along an east-west transect across Dixon Entrance, and along Portland Inlet and Portland Canal (Figures 1 and 17), clearly show that the crust-mantle boundary is deeper beneath the Coast Mountains than to the west, the Coast shear zone is a subvertical fault, and the deeper parts of the Alexander and Wrangellia terranes have seismic properties that are similar to oceanic crust.[18] The somewhat thicker crust under the Coast Mountains is consistent with isostatic principles, which predict that mountains or higher elevations should be underlain by thick crust if in gravitational equilibrium. However, the crust beneath the Coast Mountains is not much thicker than average continental crust. This combined with the geologic evidence for large amounts of exhumation suggest that significant amounts of erosion occurred as the mountains rose. The inferred oceanic crust beneath terranes west of the Coast Mountains and the subvertical boundary between the Coast Plutonic Complex and the western terranes are compatible with strike-slip accretion of island arcs west of the Coast Mountains.

Glaciers caused most of the erosion, and the resulting exhumation, in the Coast Mountains. Glaciers can erode rock more effectively than any other mechanism (for example, streams and wind) and are likely to cause some of the most rapid erosion on earth.[19] Recent studies suggest that this rapid glacial erosion of mountain belts may actually keep pace with tectonic uplift, thus limiting the height of mountain belts. Therefore, for any given climatic regime, there is likely to be a maximum mountain belt height. In Southeast Alaska, plate reconstructions indicate that plate convergence angles between the Pacific plate and North America (Figure 17), and presumably tectonic uplift, were higher during the mid-Tertiary than they are today.[20] If glacial erosion matched the tectonic uplift rates, then the Coast Mountains may not have been significantly higher than today and mountain heights may be near the maximum for the cool, moist environment.

Geologic Subdivisions

The Coast Plutonic Complex is commonly subdivided into three parallel belts of similar rocks and structures that extend the length of Southeast Alaska (Figure 17): the western metamorphic belt, the central pluton-gneiss belt, and the eastern metamorphic belt.

Rocks in the western metamorphic belt underlie the mountains along the Inside Passage (the most-used ship and Alaska Marine Highway route) from Ketchikan to Petersburg, Juneau, and Haines. The rocks have been carved into numerous fjords by glaciers that flowed westward from the higher mountains to the east. The western metamorphic belt is a sequence of volcanic arc and sedimentary rocks that formed in a marine basin 100 to 200 million years ago. The volcanic rocks probably resulted from subduction of oceanic crust similar to the way volcanism occurs in the Philippine and Japanese island arcs today. However, it is uncertain how many different island arc remnants are present or whether the subducting oceanic crust was on the east or west side of the western metamorphic belt. The Gravina belt, some of the best-known and the youngest sequence of rocks in the western metamorphic belt, crops out along its western edge. Gravina belt rocks are named for their occurrences on Gravina Island and are known to extend the length of Southeast Alaska and into the Yukon territory.[21] The belt is composed of immature sediments, including greywacke, deposited as high-density sediment flows into the marine basin[22] and island arc volcanic rocks.[23] The sediment was mostly derived from erosion of upper crustal rocks probably including nearby volcanoes within the Gravina belt. Thus, Gravina sediments were deposited in a basin that lacked sediment input from North America. Original (premetamorphic) rock types and early histories for rocks east of the Gravina belt are poorly known; an outline of current knowledge is provided in sections on geographic areas of the Coast Mountains below.

About 100 million years ago rocks in the western metamorphic belt were buried 6 to 12 miles (10 to 15 km) below the earth's surface, which caused regional metamorphism at low to medium grade (temperatures of 482 to 1,112°F or 250 to 600°C). This burial resulted from collision of the Insular and Intermontane superterranes. Paleomagnetic signatures in the rocks and fossils suggest that all of the rocks may have been much farther south than their present latitude and may not have been part of North America at that time. Folding and faulting during metamorphism tilted the sedimentary and volcanic layers and created a new metamorphic layering that is generally subvertical. The faults, including the Coast shear zone, accommodated uplift of the central pluton–gneiss belt by transporting these rocks upward and westward over the rocks to the west (Figures 17 and 18). Weathering and erosion of the faults and layering has locally produced a pronounced "topographic grain" of gullies and ridges that reflect this layering where it crosses the mountains. This can be readily observed on the hillsides east of Juneau and along the ridges above the entrance to Taku Inlet and Tracy Arm.

The western metamorphic belt is well known for mineral wealth. Copper, lead, zinc, and gold have been mined throughout the length of the belt from Ketchikan to north of Juneau. The richest gold mines are in the Juneau gold belt, which stretches from north of Juneau to the Tracy Arm and Holkham Bay area (Figures 1 and 17). The source for this mineral wealth and brief discussion of the mines are provided in the regional geology sections below.

Igneous plutons and high-temperature metamorphic rocks of the central pluton-gneiss belt, which formed more than 12 miles (19 km) below the earth's surface, underlie the highest peaks of the Coast Mountains. Some of the rocks lack pronounced layering and are very resistant to erosion; therefore, many of the mountains are smooth domes and cliffs that are steep enough and high enough to lack significant vegetation. Much of this belt is nearly inaccessible because of the high ice-covered mountains that are only locally cut by fjords and rivers. Large river systems south of Juneau (Taku), at Snettisham (Whiting), near Wrangell (Stikine), and at Prince Rupert (Skeena) form the only easy surface routes through the mountains into the interior of British Columbia. The most spectacular access to the central pluton-gneiss belt is obtained along narrow fjords, like Tracy Arm, that terminate in the high peaks of the Coast Mountains.

The central pluton-gneiss belt was formed by collision of the Insular and Intermontane superterranes, followed by magma intrusion during later subduction (Figure 18). The origin of metamorphic rocks is difficult to determine because of the high-temperature (over 1,112°F or 600°C) metamorphism, which has destroyed many of the primary features. However, limited data from metamorphosed sedimentary rocks indicate that parts of the belt are pieces of terranes currently found to the west and east that have been faulted together and heated to high temperatures. The plutons were mostly formed by Cretaceous and Tertiary magmas that rose through the amalgamation of terranes during the subduction of oceanic crust beneath the Chugach terrane to the west.

Metamorphic rocks of the eastern metamorphic belt are somewhat similar to those of the western metamorphic belt and underlie the eastern side of the Coast Mountains, in British Columbia. This nearly roadless area is still quite inaccessible because few of the waterways of Southeast Alaska extend across the high peaks along the Alaska-Canada border and only limited access is available from the Cassiar Highway in central British Columbia.

The eastern metamorphic belt is a mixture of low- to medium-grade metamorphic rocks that were juxtaposed by a series of thrust faults that transported metamorphic rocks of the Coast Plutonic Complex eastward onto the Intermontane superterrane. Large parts of the eastern metamorphic belt and the Intermontane superterrane in central British Columbia are dominated by Mesozoic volcanic rocks of the Stikine terrane.

FIGURE 19. The Fairweather Range with diorite of George Island in the foreground as seen on an unusually sunny day. The nearly three-mile-wide (4.8 km) terminus of Brady Glacier can be seen at the base of the ice- and snow-covered mountains.

CHAPTER 8

Regional Geology

Glacier Bay

T he Glacier Bay region forms the northwestern corner of Southeast Alaska (Figure 1) on the east flank of the Fairweather Range, which, along with the contiguous St. Elias Mountains, forms the highest mountains along the north Pacific coast. The highest of these mountains, Mt. Logan at about 19,000 feet (5,706 m), lies in Canada, whereas the second and third highest, Mt. St. Elias at 18,008 feet (5,408 m) and Mt. Fairweather at 15,300 feet (4,600 m), lie along the mainland strip of Alaska extending from Yakutat to west of Glacier Bay. The string of high ice-covered peaks from Mt. Fairweather to Mt. Bertha provide the source of ice for the Brady Icefield (Figure 19) and many of the glaciers that flow into Glacier Bay. Icy Strait and the southern parts of Glacier Bay are well known for abundant marine mammals during the summer months. Strong currents and upwelling water provide nutrients for the small organisms and kelp beds that are an abundant food source for whales and favored spots for sea otters.

Glacier Bay contains spectacular glaciers, glacial geomorphology, and bedrock geology. The area is currently one of only three in Southeast Alaska that has tidewater glaciers (the other two are the Tracy Arm–Endicott Arm and LeConte Bay areas). It is also home to the largest and most active glaciers in Southeast Alaska (Figures

20 and 21). These glaciers have retreated more than 60 miles in the last two hundred years, leaving behind a dense network of fjords, moraines, and other glacial features. Lateral moraines and erosional features along the fjords indicate that the ice was once over 3,000 feet thick (900 meters) and that its volume was about 550 cubic miles.[24] In spite of their dramatic retreat, many glaciers around Glacier Bay still reach the ocean.

The oral history of the indigenous Tlingit Indians describes how the Indians abandoned villages along Glacier Bay and moved to Hoonah because the glaciers were advancing across their homes.[25] This is thought to have occurred about three hundred years ago when the Little Ice Age caused widespread glacial advance. In 1794, when George Vancouver sailed into Icy Strait, ice filled Glacier Bay to the area around Bartlett Cove and the bay was only about 6 miles (9.6 km) long (Figure 21). By the time that John Muir first visited Glacier Bay around 1900, the ice had retreated to the location of Muir Point in Muir Inlet. The ice retreated an incredible 65 miles (105 km) up Muir and Tarr inlets from 1794 until its maximum retreat about 1925. Intriguingly, several of the glaciers, including Grand Pacific, advanced for several decades after 1925, but have been stagnant since about 1995.

► **FIGURE 20.** (top) The 200-foot (61-m) high vertical terminus of Margerie Glacier with a large iceberg calving into Tarr Inlet, Glacier Bay. Margerie and the adjacent Grand Pacific glaciers have retreated dramatically in historic times. However, since about 1912, these two glaciers have rejoined largely as a result of advance by the Grand Pacific Glacier. (bottom) Color infrared aerial photograph of Grand Pacific and Margerie glaciers taken from an altitude of 60,000 feet (18,293 m) (NASA Ames). The two glaciers do not appear to be joined in this image; however, today advance of the ice has rejoined them. Variation in color from white to black results from tremendous differences in the amount of rock carried by the glaciers. This image clearly shows that most of the rock debris seen from the water along the face of Grand Pacific Glacier is carried by Ferris Glacier.

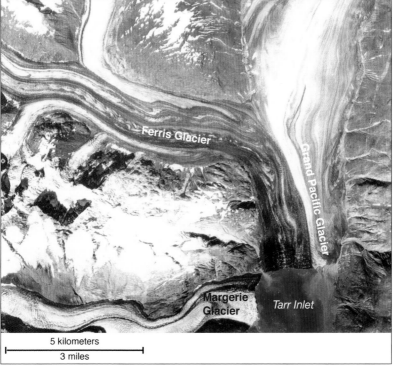

Ferris Glacier

Grand Pacific Glacier

Margerie
Glacier

Tarr Inlet

5 kilometers

3 miles

Geologists have long debated and puzzled over the causes of rapid glacial advance and retreat. By far the better data are on the rates of glacial retreat because of extensive historical records of ice retreat in Europe and North America. The rapid retreat of ice at Glacier Bay is particularly astonishing. Why did the ice retreat so fast? Was it triggered by changes in climate? Could there be other causes for retreat? It is easy to envision how an increase in mean temperature or decrease in precipitation could result in retreat of a glacier. However, in the last few decades, an alternative explanation for the rapid retreat in Glacier Bay has been proposed.

Several studies have suggested a cyclicity in tidewater glaciers that starts with slow advance of the ice as it pushes a wall of rock, or moraine, in front of it, or as glacial rivers build an outwash complex. The moraine or outwash sediments protect the wall of ice from the erosive power of saltwater. Should the glacier advance to a point where the ocean is significantly deeper (for example, Icy Strait south of Glacier Bay), the moraine may be pushed off the edge into deeper water. The glacier then "instantly" loses its protective wall and is rapidly eroded by the ocean. This initiates a rapid retreat (about 60 miles or 96 km in two hundred years in the case of Glacier Bay) that can only be slowed by retreat of the ice onto land or accumulation of enough till in front of the ice to create a new protective moraine. When accumulation of a new moraine protects the glacier, it may begin a new advance if climate is conducive.

Several glaciers around Glacier Bay have accumulated moraines and advanced southward short distances (for example, Brady, Johns Hopkins, and Grand Pacific). However, these advances have slowed

► **FIGURE 21.** Geography and geology of the Glacier Bay area showing major geologic terranes, glaciers, and lines of equal uplift (dashed yellow lines, in centimeters per year) derived from global positioning system measurements.

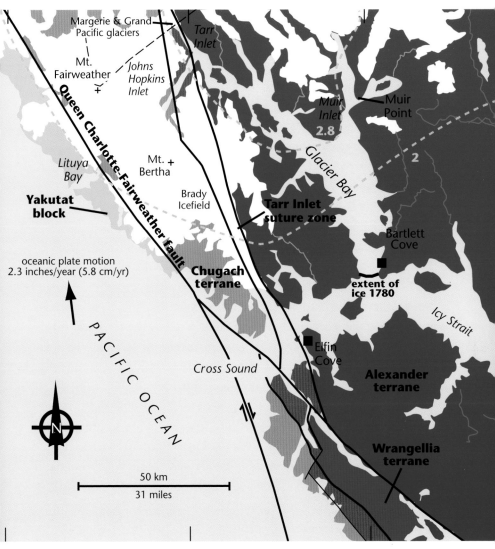

Margerie & Grand
Pacific glaciers

*Tarr
Inlet*

Mt.
Fairweather   *Johns
Hopkins
Inlet*

*Muir
Inlet*   Muir
Point

Queen Charlotte-Fairweather fault

2.8

*Lituya
Bay*   Mt. +
Bertha

Glacier Bay

**Yakutat
block**   Brady
Icefield

2

**Tarr Inlet
suture zone**

Bartlett
Cove

oceanic plate motion
2.3 inches/year (5.8 cm/yr)

**Chugach
terrane**

**extent of
ice 1780**

Icy Strait

PACIFIC OCEAN

Cross Sound   Elfin
Cove

**Alexander
terrane**

N

**Wrangellia
terrane**

50 km
31 miles

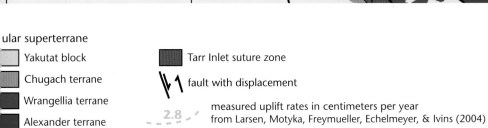

ular superterrane

- Yakutat block
- Chugach terrane
- Wrangellia terrane
- Alexander terrane

Tarr Inlet suture zone

fault with displacement

2.8   measured uplift rates in centimeters per year
from Larsen, Motyka, Freymueller, Echelmeyer, & Ivins (2004)

or halted in the last decade. Of course, it is possible, if not likely, that more than one factor may have contributed to the large-scale glacial retreat observed at Glacier Bay. For example, the long glacier that filled the bay around 1800 could have been out of equilibrium with respect to the amount of snowfall and temperature. In this case, any perturbation could cause massive collapse, and earthquake or storm energy could remove or weaken a protective moraine, triggering retreat.

One of the most remarkable aspects of the glacially carved mountains in Southeast Alaska is the rapid nature of plant succession that can form thick forests on bare rocks within a few hundred years. This is dramatically illustrated at Glacier Bay. When the first European explorers came to the bay, the area that is now Bartlett Cove and Gustavus was a large *outwash complex* composed of stream gravel that was originally deposited as the terminal moraine in front of the glacier, then reworked by meltwater. During the last two hundred years a dense forest of spruce and hemlock has grown over most of the flat area of this old outwash complex.

When glaciers filled Glacier Bay, it has been estimated that the ice was up to 3,000 feet (900 m) thick. This ice depressed the crust with its great weight. Subsequent melting dramatically lessened the weight and led to remarkable uplift or rebound of the crust. Geologists are able to measure the current uplift of the crust with precise surveying techniques and in some cases estimate former uplift rates from rocks (for example, dated marine beaches now above sea level). The Glacier Bay landscape is currently rising at rates over 1 inch (almost 3 cm) per year (Figure 21). Although it is uncertain exactly how much of this is a result of glacial rebound and how much results from tectonic forces, some researchers have concluded that these high rates of uplift cannot be solely the result of glacial rebound.

The bedrock geology of the Glacier Bay area is every bit as fascinating as the glaciers and their history. The area around Glacier Bay comprises parts of the three tectonic terranes in Southeast Alaska: the Alexander, Wrangellia, and Chugach terranes.[26] Rocks of these terranes and the faults that separate them are beautifully exposed in the sparsely vegetated mountains and fjords of the bay.

On the relatively uncommon clear days in Glacier Bay, the spectacular ice-covered mountains of the Fairweather Range dominate the northwestern skyline (Figures 3 and 19). On more typical days, when these high mountains are shrouded by clouds, the most pronounced feature seen at the mouth of the bay is a long, low tree-covered peninsula and low islands. These are remnants of the moraine and outwash complex that once lay in front of the ice that filled Glacier Bay. The low peninsula provides the sheltered anchorage, known as Bartlett Cove, for Glacier Bay Lodge. North of the islands, visitors see the low glacially carved Marble Islands within the broad waters of the lower bay. These rocks and other carbonate rocks found along the eastern and western shores of Glacier Bay are a fault slice of deformed and somewhat metamorphosed lower Paleozoic limestone from the Alexander terrane, which is much less deformed and better exposed to the south in the Alexander Archipelago. These carbonate rocks and other associated sediments provide evidence for shallow marine deposition for at least part of the Alexander terrane during the Silurian and Devonian. Northwest of the entrance to Glacier Bay, in Johns Hopkins Inlet, is perhaps the best-exposed example of faults that juxtapose the Paleozoic Alexander terrane and Cretaceous Chugach terrane. Here, faults that are well exposed on the north and south side of the inlet have broken up and displaced large (up to about a mile in size) pieces of rock creating a tectonic mélange. "Mélange" is a French word used to describe rocks that have been tectonically broken up and mixed. This fault zone is known as the

Tarr Inlet *suture zone*. At the northern extremity of the bay, Tarr Inlet is largely surrounded by mountains sculpted in a Cretaceous igneous rock called *granodiorite*. The solid granodiorite is partially obscured by thin benches of lateral moraine and *alluvial fans* of reworked till.

The Glacier Bay area has not only ancient sutures and faults, but also currently active faults. The present movement of the Pacific plate is northward with respect to the North American plate. Motion between the two plates is accommodated by the Queen Charlotte-Fairweather fault that subparallels the Pacific coast west of Southeast Alaska (Figure 17) and extends onto land within the western part of Glacier Bay National Park and Preserve (Figure 21). Although movement of the Pacific plate is dominantly northward, a component of compression across the fault is causing uplift of the Fairweather and St. Elias mountains. The Queen Charlotte-Fairweather fault has probably accommodated movement between the Pacific and North American plates for the last 35 million years. For the last few thousand years, the rate of horizontal movement has been estimated at 2 to 4 inches (5 to 10 cm) per year based on offsets of stream valleys dated at less than thirteen hundred years old and GPS measurements.[27] Simultaneous uplift of the Fairweather mountains is estimated to be about $\frac{1}{16}$ inch per year (1.5 mm/yr).[28]

Movement along the Queen Charlotte–Fairweather fault has caused extensive earthquake activity. Although minor quakes are common along the coast of Alaska, there is often little damage to land and people. A notable exception occurred in 1958 when a major earthquake caused about 40 million cubic yards (30 million m^3) of rock to slide 3,000 feet (900 m) downslope into Lituya Bay (west of Glacier Bay on the Gulf of Alaska, Figure 21). The resulting 100-foot (30-m) wave crossed the bay at about 100 miles per hour (160 km/hr) and traveled to a height of 1,720 feet (517 m) on the opposite side of the bay. At the time of the earthquake, there were three fishing boats in the bay as the wave formed. Amazingly, one of the boats survived the towering wave without being swamped. Unfortunately, the other two boats sank and two people were killed. The wave destroyed about 4 square miles (10 km²) of trees below an elevation of 1,800 feet (540 m) on the shore of the bay.

Haines and Skagway

Haines and Skagway lie at the northeastern corner of Southeast Alaska, only a few miles from the Canadian border (Figure 1). Although both are small towns perched on the edge of steep fjords, they have had a long-term role as important ports that provide routes to the Yukon Territory and Interior Alaska. Haines, on the northwest side of Chilkoot Inlet, provides convenient access to the Alaska-Canada Highway through Haines Junction. Skagway, on the shore of steep Taiya Inlet, provides access to the Yukon through Whitehorse. These two narrow inlets branch from the northern end of the long and spectacular fjord of Lynn Canal.

Travelers on the Inside Passage going to Haines or Skagway along Lynn Canal see some of the most spectacular evidence of glaciation and mountain building. North of Juneau, several large rivers dissect the Coast Mountains and feed large amounts of glacial silt into the waters of Berners Bay (Figure 1). Fine silt suspended in the river water produces a beautiful blue-green color that typically forms abrupt lines separating it from the darker ocean water. Mountains surrounding the bay are Gravina belt volcanic basalt and sediments intruded by a pluton of granodiorite just north of the bay. Quartz veins in the pluton provided gold for the northernmost mines in the Juneau gold belt.

Further north, Lynn Canal is a long, straight fjord following the trace of the Chatham Strait fault on an oblique route through the Coast Mountains. The Chatham Strait fault (Figure 17) is a strike-slip fault that transported rocks on the west northward and connects with the Denali fault in Southcentral Alaska. On the west, Alexander terrane rocks of the Chilkat Mountains form snow-capped peaks that lie east of Glacier Bay. On the east, the central pluton-gneiss belt separates Southeast Alaska from British Columbia. Lynn Canal ends in spectacular cliff exposures of Late Cretaceous to Paleocene *tonalite* and granodiorite near Skagway. The coastline adjacent to Haines is underlain by low-grade metamorphic rocks of the Wrangellia terrane that are juxtaposed with rocks of the Coast Plutonic Complex to the east.[29]

The Klondike Gold Rush

Interior Alaska and the Yukon Territory were largely settled and explored as a result of intrepid gold seekers who braved incredible hardships in order to seek their fortunes. In 1896, Skookum Jim Mason, Dawson Charlie, and George Washington Carmack discovered gold in the Klondike River of the Yukon Territory. By 1898, the discoveries set off a gold rush that inspired thousands of fortune seekers to head north for the Klondike.

Unfortunately, the riches were not readily found and there was no easy route to the supposed gold fields. Participants in the Klondike-Yukon gold rush first had to travel by ship up the Inside Passage, then after landing at Skagway, they had to climb the icy 3,739-foot (1,123-m) high Chilkoot Pass in order to cross into Canada and finish the 600-mile (960-km) trek to the Klondike. Many fortune seekers died during the cold winters of 1898 and 1899. Word of physical hardships and a lack of ready fortunes in the gold fields rapidly reached the cities in the south and fortune seekers ceased coming north. However, Skagway, and now Haines with construction of the Haines Highway, still provides access to the wonders of northern Canada and the interior of Alaska.

Sitka

Sitka lies on the western coast of Baranof Island (Figure 1), with only a few small islands between the city and the Pacific Ocean. Although Sitka Harbor has easy access to the Gulf of Alaska, small boats prefer to arrive through the narrow and protected Peril Strait waterway, which separates Chichagof and Baranof islands (Figures 1 and 22). The city is only about 10 miles (16 km) east of Mt. Edgecumbe and Crater Ridge on Kruzof Island—the only Holocene volcanic field in Southeast Alaska. However, the city and much of Baranof Island are underlain by immature clastic sediments, volcanic rocks, and mafic igneous plutons of the Chugach terrane (Figure 22). These rocks were scraped from a subducting oceanic plate into an accretionary prism about 100 million years ago. As the prism thickened and additional rocks were added, they were faulted and deformed into a tectonic mixture or mélange. Sedimentary rocks from this area and the Gravina belt to the east are some of the youngest sedimentary rocks in Southeast Alaska. Cretaceous and Tertiary plutons of tonalite and granodiorite intruded broad areas of the Chugach terrane south and east of Sitka (Figure 22). Many examples of hot springs can be found on Baranof and Chichagof Islands. In these areas, hot volcanic and plutonic igneous rocks combined with faults allow water to circulate downward, heat up, and then rise to form hot springs. Examples of these hot springs are found around the town of Baranof on Baranof Island (Figure 22).

Sitka is the oldest non-Native settlement in Southeast Alaska. The city was founded when Alexander Baranof, chief governor of the Russian-American Company, moved the company headquarters

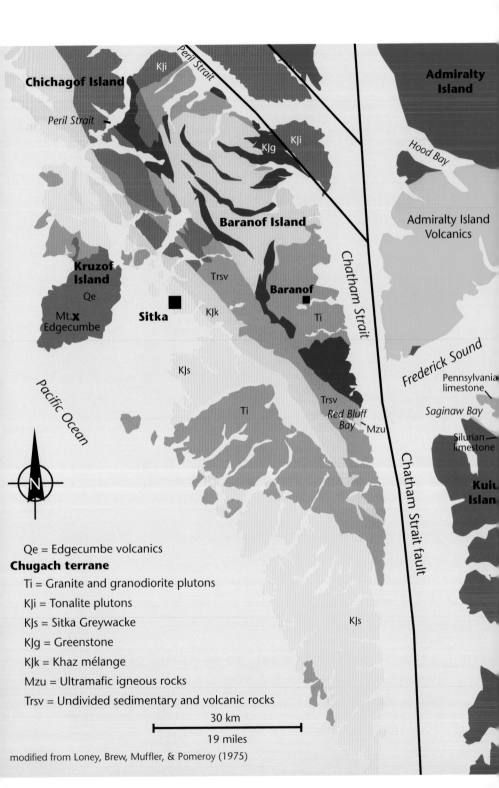

Chichagof Island

Peril Strait

Peril Strait

KJi

KJg

KJi

Admiralty Island

Hood Bay

Baranof Island

Admiralty Island Volcanics

Kruzof Island

Qe

Mt. **X** Edgecumbe

Sitka

Trsv

KJk

Baranof

Ti

Chatham Strait

Pacific Ocean

KJs

Ti

Trsv

Red Bluff Bay

Mzu

Frederick Sound

Pennsylvania limestone

Saginaw Bay

Silurian limestone

Chatham Strait fault

Kuiu Islan

N

KJs

Qe = Edgecumbe volcanics

Chugach terrane

Ti = Granite and granodiorite plutons

KJi = Tonalite plutons

KJs = Sitka Greywacke

KJg = Greenstone

KJk = Khaz mélange

Mzu = Ultramafic igneous rocks

Trsv = Undivided sedimentary and volcanic rocks

30 km

19 miles

modified from Loney, Brew, Muffler, & Pomeroy (1975)

there from Kodiak in 1799. In 1804, Sitka became the capital of the Russian territory and it remained the capital of Russian Alaska until the purchase of Alaska by the United States in 1867.

Sitka, like Juneau, has a long history of mining for precious metals. The first gold discoveries near Sitka were about 1871. *Lode gold*, ore found in solid rock, was mined and milled at Stewart Ledge on Silver Bay southeast of the town. Although few large deposits were discovered near Sitka, the second richest gold mines in Southeast Alaska were in the Chichagof district on the western side of Chichagof Island, north of Sitka. These deposits were mined until World War II, when labor shortages forced closure of the mines. Gold mineralization occurred in numerous fractures that allowed hot fluids to percolate upward and cool. The cooling fluids deposited gold and quartz in a network of veins similar to the most productive mines in the Juneau gold belt.

The west side of Chichagof Island and most of Baranof Island are underlain by weakly metamorphosed but strongly deformed rocks that were accreted to the older Wrangell and Alexander terrane rocks to the north and east during late Cretaceous and early Tertiary subduction of the Pacific plate beneath North America (Figures 17 and 18). These Chugach terrane rocks (Figure 17) are largely mélange of the Kelp Bay Group and sedimentary rocks known as the Sitka Greywacke. The Kelp Bay Group is a mixture of metamorphosed late Mesozoic volcanic and sedimentary rocks that were tectonically mixed during subduction beneath the Chugach terrane. "Greywacke" is a term used to describe sedimentary rock made of unsorted fragments or clasts of rocks and minerals in a matrix of

◄ **FIGURE 22.** Simplified geologic map of Baranof Island and adjacent parts of Admiralty, Chichagof, and Kuiu islands. Most of the rocks that underlie Baranof Island are part of the Chugach terrane (rocks below Chugach in legend) that formed as an accretionary prism above east-dipping subduction of oceanic crust beneath the cordillera during the late Mesozoic.

mud. The Sitka Greywacke was deposited along the edge of North America during the Cretaceous when dense sediment-laden bottom currents carried materials into the adjacent ocean. These rocks form the rocky headlands along the beach south of Sitka.

Mt. Edgecumbe (Figure 23) is a prominent 3,201-foot (961-m) composite volcano that can be seen from Sitka. This volcano and the adjacent collapsed caldera of Crater Ridge are parts of the northeast-trending 100-square-mile (256-km^2) Mt. Edgecumbe volcanic field. Although some of the basalt, basaltic andesite, andesite, and rhyolite is up to several million years old, the last volcanic eruptions that produced ash deposits occurred about 11,000 and 4,500 years ago,[30] and there has been no activity of any type in the last two hundred years. This volcanism is not directly related to subduction because the nearby North America and Pacific plate boundary (Figure 16) is one of strike-slip motion. Rather, the volcanism may be related to magma migration up this boundary, which is formed by the Queen Charlotte-Fairweather fault.

Mt. Edgecumbe was the site of a rather dramatic April Fools' joke. In 1974, residents of Sitka noticed a plume of smoke coming from the crater at the top of the volcano. The "eruption" turned out to

FIGURE 23. Mt. Edgecumbe composite volcano viewed looking west from Baranof Island. The summit is 3,201 feet (976 m) above sea level.

be a large pile of old tires, placed in the small crater at the top of the volcano by a group of locals using a helicopter, and set on fire on April 1.

Chatham Strait

The prominent linear waterway separating Baranof and Chichagof islands on the west from Kuiu and Admiralty islands to the east is known as Chatham Strait (Figure 1). The width and orientation of the strait allow southwest winds to kick up significant waves, making it one of the roughest passageways between the Southeast Alaska islands. Fortunately for boaters, numerous narrow fjords lead from the strait into Baranof Island providing small but protected anchorages.

The straight character of this waterway results directly from the underlying geology. A massive glacier eroded the channel as it followed the path of one of the major faults in the region: the Chatham Strait fault. Lateral motion along the fault moved rocks on the west side about 100 miles (160 km) northward with respect to those on the east side.[31] The eastern shores of Baranof and Chichagof islands and snow-capped mountains beyond are underlain by rocks of the Chugach terrane. At Red Bluff Bay (Figures 22 and 24), magnesium-rich *ultramafic* rocks form barren red hillsides owing to the poor soils formed from these igneous rocks. These ultramafic rocks are often called "Alpine-type ultramafics" for the Alps where the rock type was first described. The origin of these bodies is hard to determine because deformation and metamorphism obscure any primary features. To the north, Chugach terrane rocks are intruded by a large pluton of granite that is well exposed around the waterfall at Baranof (Figure 22). The more subdued slopes on the western shores of Kuiu and Admiralty islands are dominated by sedimentary rocks of the Alexander terrane locally overlain by younger volcanic flows discussed in the section below on Frederick Sound.

FIGURE 24. The red hillside on the north shore of Red Bluff Bay provides a dramatic entrance to this short but beautiful fjord on the eastern side of Baranof Island. The red color is formed by oxidation of the magnesium and iron-rich ultramafic igneous rock. Chemical weathering of this rock produces poor soils and stunts the growth of trees, which never obtain large size and look much like bonsai trees.

Juneau

The capital city of Alaska is situated on either side of the narrow and linear water-filled Gastineau Channel, between the mountains of Douglas Island on the west and the mainland on the east (Figure 1). High mountains east of the city form a rock wall between Gastineau Channel and the Juneau Icefield, one of the largest icefields in the Pacific Northwest with an area of over 1,200 square miles (3,070 km²). Within a mile of downtown, mountains rise to elevations of more than 4,000 feet (1,200 m) and the higher peaks within the Juneau Icefield are more than 7,000 feet (2,100 m) high. These mountains cause warm moisture-laden air from the Gulf of Alaska to rise and drop moisture that feeds the icefield, which in turn feeds glaciers that flow outward in all directions (Figure 25). Most of the glaciers from the icefield are retreating and none of the ice currently reaches the ocean.

Downtown Juneau is rich in historical buildings and the remains of old mine workings. However, old businesses and local character have been largely eliminated or modified by the huge influence of state government buildings and businesses that cater to the cruise ships that overrun the city during the summer. In fact, the changes are so overwhelming that during summer many residents spend much of their time in the calm of Mendenhall Valley and other points north of downtown.

The glaciers that flow from the Juneau Icefield have retreated dramatically since Juneau was established. Taku Glacier on Taku Inlet briefly blocked the Taku River in 1750 and was a tidewater glacier until about 1900, when it retreated enough for a large tidal flat to form in front of the ice. Although all other glaciers in the Juneau Icefield are retreating, the Taku and Hole-in-the-Wall glaciers, north of Taku River, are currently advancing. The advance of these glaciers is a serious concern to planners who have proposed building a road along the Taku River from Juneau into British Columbia. Seismological studies on the Taku Glacier indicate that it is the thickest temperate glacier known with ice more than 4,600 feet (1,402 m) thick.[32] Its terminus is protected by thick accumulation of sediment largely deposited by the sediment-laden Taku River.

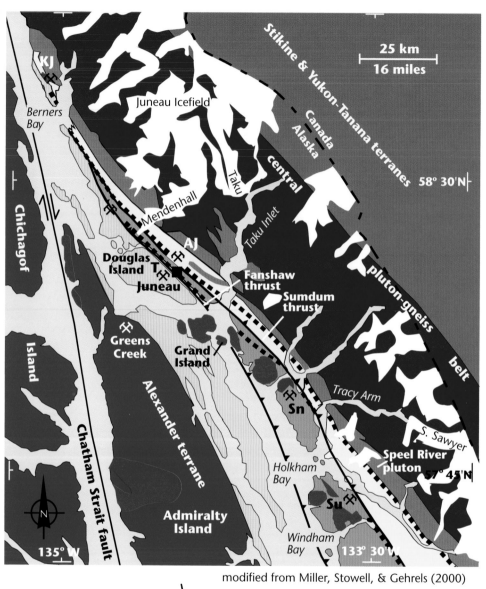

modified from Miller, Stowell, & Gehrels (2000)

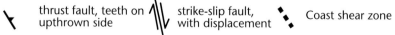

⚹ thrust fault, teeth on upthrown side　⚹⚹ strike-slip fault, with displacement　▪▪▪ Coast shear zone

✠ Gold districts: KJ = Kensington-Jualin, AJ = Alaska-Juneau, T = Treadwell, Sn = Snettisham, Su = Sumdum

western metamorphic belt

Gravina belt

Taku terrane

Yukon-Tanana terrane

Coast Plutonic Complex sill plutons

mid-Cretaceous plutons

Alexander terrane

Downtown Juneau is nestled in the narrowest part of Gastineau Channel near two glacial valleys (Gold Creek and Sheep Creek valleys) that lead eastward up to rich gold mines responsible for the city's location. This section of the Gastineau Channel is so narrow that much of the town was built on mine tailings that were dumped into the marine channel at the base of the steep avalanche-scarred slopes of Mt. Roberts and Mt. Juneau, directly east of town. The northern part of Juneau, including the airport and most of the suburbs, occupies the flat-bottomed Mendenhall Valley, which was vacated as the Mendenhall Glacier retreated to the east. The valley is floored by post–ice age marine sediments overlain by glacial outwash and mud, and in the upper valley glacial moraines from the Little Ice Age retreat of Mendenhall Glacier. The glacier has retreated about 3 miles (4.8 km) since 1767, providing space for a small lake, the National Forest Visitors Center, and a campground. The sediment-laden Mendenhall River has modified the valley and nearly blocked the northern part of Gastineau Channel with a wide delta since retreat of the glacier.

Juneau sits astride the boundary between low-temperature metamorphic rocks of the Gravina belt on the west (Douglas Island) and higher-temperature metamorphic rocks of the Taku terrane on the east (mainland). Taku terrane rocks were intruded by 50-to-70-million-year-old tonalite and granite of the central pluton-gneiss belt; these hard rocks underlie high peaks of the Juneau Icefield, east of town.

The Gastineau Channel is one of the most impressive of the linear glacially carved channels that parallel the Coast Mountains between Lynn Canal on the north and Prince Rupert on the south. These long depressions, or *lineaments*, mark the location of

◄ **FIGURE 25.** Geography, geology, and gold mines of the Juneau gold belt. Icefields and glaciers are only drawn up to the Canadian border. Only the most important gold mines and districts are shown.

numerous faults collectively called the Coast shear zone (Figures 17 and 25). Most of the displacements probably occurred between 90 and 50 million years ago; however, some movement on the Gastineau Channel fault is more recent. Faulting along Gastineau Channel juxtaposed older Taku terrane and younger Gravina belt rocks. The steep-sided channel, which provides the dramatic setting for Juneau, formed when glaciers preferentially carved the less resistant faulted and sheared rocks of the Coast shear zone.

Gold Mining and the Westernization of Southeast Alaska

The earliest European settlers in Southeast Alaska were explorers and fur traders from Russia. Although early Russian traders and hunters fought bloody conflicts with the native Indian tribes, the Russians and later American hunters did not occupy much of the land. This sparse settlement by Europeans changed with the discovery of gold in 1870. Gold was first found at Sumdum and Windham Bay, about 60 miles (100 km) south of Juneau. Six major gold districts were developed in what is now called the Juneau gold belt. Shortly after the first discovery, Chief Kowee led Joe Juneau and Richard Harris to *placer gold* deposits (gold-rich stream sediments) in Silverbow Basin about 2 miles (3.2 km) east of Juneau in Gold Creek. Prospectors soon found lode or hard-rock gold veins above the stream deposits in Silverbow Basin.

The gold discoveries in Silverbow Basin and later finds along the Treadwell trend of mines on Douglas Island proved to be the largest deposits found to date in Southeast Alaska. This area of the Coast Mountains between Berners Bay and Windham Bay was soon named the Juneau gold belt (Figure 25). Gold deposits in this belt provided impetus for development of the three largest mines: the Treadwell on Douglas Island and the Alaska-Gastineau and Alaska-

Juneau on the mainland east of Juneau. At the peak of production, these three mines were processing a total of about 30,000 tons of ore a day and employing thousands of miners. The mines eventually produced nearly seven million ounces of gold ($2.1 billion at $300 per ounce). The Treadwell mine operated until 1917, when removal of pillars holding the roof apparently caused collapse of the roof and flooding of the mine, which extended more than 1,500 feet (457 m) below Gastineau Channel. Production continued in the Alaska-Juneau mine until 1944, when labor shortages forced closure due to World War II. The mining led to tremendous economic growth and engineering and building developments: housing and recreation facilities were constructed; workers were brought in from China and the Philippines; hydroelectric power stations, including a concrete compression arch dam with dimensions of 650 by 175 feet (198 x 53 m), were built to power the mines; and areas of Gastineau Channel were filled with mine tailings to build downtown Juneau.[33]

Extensive information about the Juneau gold belt and other mineral deposits in Southeast Alaska can be obtained from the Juneau Mineral Information Center in Douglas (Bureau of Land Management). Some of this information is available online.

Southeast Alaska has seen considerable precious metals exploration during the last twenty years. Much of this exploration focused on reopening old mines; few new mines have resulted. Feasibility studies have been done to reopen the Kensington mine north of Berners Bay and the Alaska-Juneau mine near Juneau. However, the Alaska-Juneau project was placed on long-term hold in 1997 because of the low price of gold and problems with obtaining approval for mine and waste-disposal plans. In 1989, a new gold, silver, and base-metal mine (Greens Creek) opened on the northern end of Admiralty Island (Figure 25).

Cyclic Garnet Zoning: A Record of Fluid Flow in the Crust

Chemical zoning of minerals can provide a record of rock history if it is preserved. Garnet and certain other minerals incorporate several chemical elements into their structure, depending on pressure, temperature, and the chemistry of the rock (available elements). As garnet crystal growth proceeds, the chemistry of the current growth increment reflects these parameters. Therefore, garnet may record the history of growth in concentric chemical layers from the center of a crystal outward. This chemical zoning may be a simple increase or decrease in concentration or more complicated cyclic variation in chemistry depending on the cause. Garnet often preserves chemical zoning because diffusion does not readily change the chemistry after growth. Therefore, the chemical zoning can be interpreted somewhat like tree rings, with interior chemistry representing a record of early growth and outer chemistry representing later growth.

Rocks may undergo thermally driven fluid flow when intrusion of hot magma or faulting produces a thermal gradient and fluid is able to flow through the rocks. Faulted rocks typically allow significant fluid flow (they are highly permeable), providing excellent pathways for fluid flow carrying elements that may be incorporated into garnet crystals.

Strongly zoned garnet crystals from the Juneau gold belt on northeast Admiralty Island preserve the history of igneous intrusion and fluid flow. These crystals show pronounced cyclic zoning of calcium and iron. Garnet growth was caused by heating from the adjacent 90-million-year-old pluton at Grand Island (Figure 25). The chemical zoning was interpreted to result from at least three pulses of fluid flow during mineral growth.[34]

BACKSCATTERED ELECTRON IMAGE of a 0.5 mm garnet crystal from near Grand Island. Colors indicate the relative concentrations of calcium and iron: blue and green indicate low iron and high calcium, red indicates high iron and low calcium.

The nature and cause of gold mineralization in the Juneau gold belt has long been a subject of intense interest. Most of the gold occurs with sulfide minerals (*pyrite* and others) in quartz plus-or-minus *calcite* mineral-filled fractures or veins. These veins cut across metamorphic or igneous rocks and therefore are younger than these hosts. Some of the earliest workers assumed that gold was derived from the numerous igneous intrusive rocks that host some of the veins. However, the most recent theories for gold deposition[35] suggest that hot water derived from dehydration of rocks during metamorphism transported gold. This water-leached gold from the metamorphic rocks then migrated upward along faults when the stress regime allowed expulsion of the fluids. Gold was deposited in open fractures forming veins (Figure 26) as the

FIGURE 26. Gold-bearing quartz and pyrite veins in the Kensington mine, north of Juneau (see Figure 25). The curved white quartz veins with gold-colored sulfide minerals are fractures that opened up, then filled with quartz + pyrite + gold as metal-rich fluids migrated through the host quartz diorite pluton. Although the veins are rich in gold, none is visible at this scale.

fluids cooled in the mid to upper crust. Incredibly, all of the gold deposits that have been dated were formed within a one-to-two-million-year period, suggesting a very rapid geologic event, such as release of stress by changes in plate motion.

Tracy Arm, Endicott Arm, and Holkham Bay

Tracy Arm and Endicott Arm are narrow fjords leading into Holkham Bay about 50 miles (80 km) southeast of Juneau (Figure 1). The two fjords extend more than 20 miles (32 km) from the western edge of the Coast Mountains eastward to tidewater glaciers surrounded by imposing mountains that rise about 7,000 feet (2,100 m) above the sea. Numerous steep valleys on either side of these fjords were once filled with ice feeding into the larger glaciers that carved the fiords (Figure 27). The steep mountainous region

FIGURE 27. U-shaped valley carved by glacial ice, south shore of Tracy Arm. The nearly flat floor of this valley is over 600 feet (183 m) above the bottom of the fjord, indicating that the erosive power of Sawyer Glacier that carved Tracy Arm was far greater than the erosive power of the tributary that carved this U-shaped valley.

from the eastern shoreline of Holkham Bay to the Canadian border has been protected as the Tracy Arm–Fords Terror Wilderness.

Fords Terror, a narrow fjord that extends eastward from Endicott Arm (Figure 28), is named for one of the early explorers who ventured up the fjord at slack tide. He and his crew, in longboats, did not realize that the extremely shallow channel at its narrowest point combined with the deep water in the over five miles of fjord beyond set the stage for a tremendous tidal bore during incoming and outgoing tides. The longboats entered at slack tide without difficulty; however, the tide was falling as they rowed out across the narrow constricted part of the fjord. The boatmen were suddenly faced by standing waves over 3 feet (1 m) in height and a curve in the channel that threatens to slam boats into a vertical wall of rock.

Spectacular evidence for the glacial processes that carved the fjords can be found throughout the Holkham Bay, Tracy Arm, and

FIGURE 28. Zodiac inflatable boat braving the standing waves in whitewater of the tidal bore near the entrance to Fords Terror (see Figure 29). Many narrow inlets, like Fords Terror, present a challenge to boats when the tide is changing.

Endicott Arm areas. The shallow water that extends across the middle of Holkham Bay barely covers the end moraines of the glaciers that carved the two fjords (Figure 29). On the north side of the bay, till has been breached by meltwater or tides to form a narrow channel into the entrance of Tracy Arm. Water rushing through this channel creates large standing waves as the tide pours in and out.

A dense forest of spruce, hemlock, and lesser amounts of cedar cling to the steep slopes underlain by the metamorphic rocks at the entrance to Tracy Arm. The nearly vertical bare rock faces east of the entrance are carved into igneous plutons and high-temperature metamorphic gneiss that is more resistant to the forces of erosion. These resistant rocks preserve wonderful examples of the information that geologists use to interpret the landscape as glacially carved. Spectacular horizontal glacial grooves have been cut into the solid rock along the walls of the fjord. U-shaped valleys can be observed at waterline or up the steep slopes along both sides. These valleys are all *hanging valleys* with respect to the bottom of the fjord, which is more than 1,000 feet (300 m) below sea level. Arêtes that once separated the flowing tributaries of ice that carved the valleys now separate some of the valleys. Pointed mountain peaks or horns can also be seen; these mountains were formed at the head of valley glaciers that were flowing in divergent directions away from the peak.

One of the most remarkable aspects of the steep glacially carved fjords in Southeast Alaska is the ability of plants to rapidly

► **FIGURE 29.** Simplified geologic map of the Holkham Bay and Tracy Arm area. Mid Cretaceous plutons are shown in medium red, Gravina belt rocks are shown in tan, the western metamorphic belt is shown in blue, and the central pluton-gneiss belt rocks are shown in dark red. Lines labeled with mineral names show the extent of metamorphic mineral growth. Modified from Stowell and Hooper (1990).

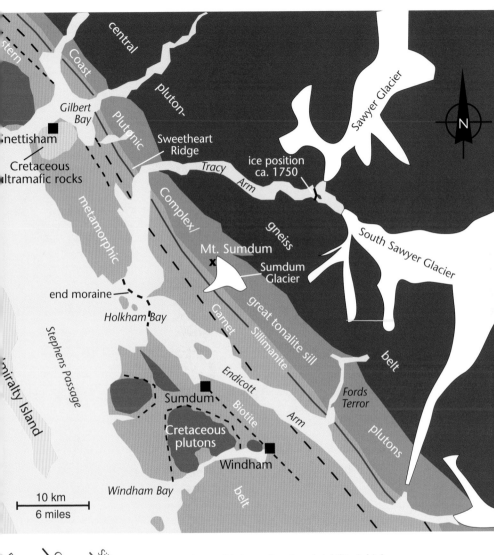

central Coast pluton-

Gilbert Bay

central

Sawyer Glacier

N

nettisham

Cretaceous ltramafic rocks

Plutonic

Sweetheart Ridge

Tracy Arm

ice position ca. 1750

metamorphic

Complex/

gneiss

Mt. Sumdum
x

Sumdum Glacier

South Sawyer Glacier

end moraine

Holkham Bay

Stephens Passage

Garnet

great tonalite sill

Sillimanite

belt

Endicott

Fords Terror

miralty Island

Sumdum ■

Biotite

Arm

plutons

Cretaceous plutons

Windham ■

10 km

6 miles

Windham Bay

belt

Biotite   Garnet   Sillimanite

metamorphic isograds: mineral stability & high temperature on the named side

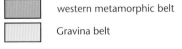

western metamorphic belt

Gravina belt

Coast Plutonic Complex sill plutons

mid-Cretaceous plutons

populate the cliffs that are initially devoid of soil. The succession of plants typically begins with moss, small flowers, and alders that can grow without much soil in small fractures in the rock or where small pockets of glacial till were left behind by the retreating ice. These plants rapidly break down some of the rock and add nutrients, allowing blueberries, devil's club, and Sitka spruce to grow. Remarkable "gardens" of all these plants can be seen clinging to the near-vertical walls of Tracy Arm.

Tracy Arm terminates in two looming walls of ice that extend about 500 feet (150 m) from the bottom of the fjord to the top of the glaciers: Sawyer and South Sawyer glaciers (Figure 29). These glaciers are very active and *calve* frequently, often choking the fjord with everything from bergy bits to large icebergs. Intriguingly, these glaciers have been far more stable than those in Glacier Bay. Although few records exist for the position of the ice in the past, charts show that Sawyer Glacier may not have retreated more than about 5 miles (8 km) in the last 150 years (Figure 29).

About 3 miles (4.8 km) from Sawyer Glacier a low island of glacially carved rock has been exposed by retreating ice within the last 150 years. Although several small to medium-sized Sitka spruce grow there, much of the island is free of vegetation. This somewhat desolate and very cold island is known for having had one of the very few known bald eagles' nests that was built on the ground. Almost all eagles build their nests in trees. In fact, another bald eagle's nest sits in a spruce tree on the mainland not more than 200 yards (183 m) from the island. In 1984, while researching the geology of Tracy Arm, I camped along the fjord with a field assistant for several nights. We spent one rather cold and wet night on the island and then walked up onto the top of the hill. As I approached the top of the hill, I suddenly heard small birds. I was shocked to see a large nest on the ground with two eagle chicks and an unhatched egg. I quickly retreated so as to avoid disturbing the birds or inviting the wrath of the mother eagle, which could have been nearby.

Rocks in the Holkham Bay area preserve a fascinating history, which started with near-surface sedimentation and volcanism more than 250 million years ago and continued sporadically until about 110 million years ago. This was followed by deep burial, metamorphism, and intrusion of plutons from about 90 to 60 million years ago. Finally, rapid uplift accompanied by erosion and cooling brought lower crust to the surface less than 60 million years ago.[36]

The western region of Holkham Bay and the entrance to Tracy and Endicott arms are underlain by the metamorphosed sedimentary and volcanic rocks of the western metamorphic belt, intruded by a few 90-million-year-old diorite plutons (Figure 29). The metamorphic rocks include *phyllite*, slate, *marble*, and greenstone. Marine fossils in the marble and ripple marks preserved in the slate indicate that these rocks were originally formed as marine sediments. The greenstone contains *pillow texture* and large *relict* feldspar crystals, suggesting that it was erupted into water as a basaltic volcanic rock. Fossils indicate that these rocks formed about 200 million years ago. Metamorphosed conglomerate found along the western shore of Gilbert Bay and near Prospect Creek contains pebbles that are more than one billion years old. Because no rocks this old are found nearby, these pebbles suggest that rivers flowing from the ancient core of North America (the Canadian Shield) deposited them. Metamorphic minerals in the rocks indicate that deep burial and heating occurred after deposition. The common metamorphic minerals, *mica*, *chlorite*, and *amphibole*, indicate that these are low-temperature metamorphic rocks, which likely formed at 570 to 750°F (300 to 400°C) and pressures corresponding to depths of 9 to 12 miles (15 to 20 km) beneath the surface. Rocks in the western metamorphic belt are intensely folded and faulted (Figure 30).

Geochronology and Zircon Crystals: Clues for Deciphering Earth History

Geochronology is the science of determining the age of earth events. Almost all modern age determinations rely on radioactive decay and some of the more refined methods use decay of elements that are incorporated into single crystals of minerals such as garnet or *zircon*.

Zircon is a zirconium silicate mineral ($ZrSiO_4$) that incorporates significant amounts of uranium into its structure. Uranium in the crystals undergoes slow radioactive decay to form lead. Because lead is not included in zircon, the proportion of uranium and the lead produced from decay provides a means for determining the time of crystal growth. Many igneous rocks contain zircon. Thus, sedimentary rocks derived by weathering them may also contain zircon because it is resistant to weathering. In fact, this mineral can survive incorporation into an igneous melt from an older rock, weathering at the earth's surface, and subsequent metamorphism, and still retain isotopic ratios that reflect the early events.

Plutons that intrude across faults clearly indicate that the faulting is older than the igneous intrusion. Therefore, a pluton age provides a minimum age for fault movement and in some cases terrane emplacement along a fault. Igneous and metamorphic rocks along the western flank of the Coast Mountains provide numerous samples for age determination. George Gehrels (University of Arizona) and colleagues have extracted zircon from a wide range of rocks and pieced together a history for pluton emplacement that provides constraints on the nature and age of the crust, and the timing of terrane accretion.[37]

For example, diorite plutons that underlie northeast Admiralty Island, Grand Island, and the mainland nearby intruded across the Fanshaw thrust (Figure 25). Thus the 94-million-year age for the pluton[38] constrains fault movement to more than 94 million years ago.

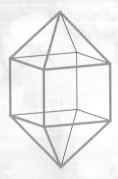

DRAWING of an idealized zircon crystal with doubly tipped elongate shape.

Eastward across Holkham Bay, the mineralogy of the rocks changes with addition of garnet and *sillimanite* and loss of chlorite, indicating temperatures of more than 1,112°F (600°C), greater than temperatures on the west side at similar depths. Little is known about the pre-metamorphic history of these rocks; however, rare ripple marks and partly recrystallized marine fossils found along the north shore of Tracy Arm indicate shallow marine deposition for some of the rocks, and rock fragments thrown from volcanoes indicate a volcanic origin for other parts of this sequence. Lack

FIGURE 30. Complex folding of low- to medium-grade slate (metamorphosed sediments) of the western metamorphic belt, Marmion Island south of Juneau.

of limestone and billion-year-old pebbles on the eastern side of Gilbert Bay and variations in the chemistry of metamorphosed volcanic rocks[39] suggest that rocks exposed along the eastern part of Holkham Bay may be far older and unrelated to the rocks on the west.

Rocks throughout the western metamorphic belt are strongly deformed into tight folds, shear zones, and faults. This deformation rotated initially horizontal sedimentary bedding and volcanic layering into a nearly vertical orientation. Intense deformation is inferred to have thrust once hot metamorphic and plutonic rocks to the east over cooler metamorphic rocks to the west.

The boundary between the western metamorphic belt and rocks of the central pluton-gneiss belt to the east lies at the first sharp bend in Tracy Arm (Figure 29). The sharp break in rock type forms an obvious break in topography and vegetation. The rounded mountains west of the contact, underlain by metamorphic rocks, are covered by a dense forest of western hemlock and Sitka spruce; however, east of the contact sheer cliffs of tonalite and gneiss are only locally cloaked with spruce, alder, or devil's club. This dramatic change results from the recent glaciation toward the east, and perhaps more importantly from the tremendous resistance of the tonalite and gneiss to erosion. The first rock type found along the western edge of the central pluton-gneiss belt is tonalite of the Coast Plutonic Complex sill.[40] This remarkably uniform series of tonalite intrusions is 6 to 9 miles (10 to 15 km) wide and stretches over 400 miles (640 km) from near Prince Rupert, British Columbia, to Skagway, Alaska. East of the Coast Plutonic Complex sill, numerous granite plutons intruded the high-temperature metamorphic rocks. Much of this gneiss reached temperatures high enough for part of it to melt. The resulting mixtures of now solidified liquid and solid metamorphic rock are called *migmatites* (Figure 31).

FIGURE 31. Intensely folded high-grade gneiss from the central pluton-gneiss belt. Rocks like these have reached metamorphic temperatures that caused partial melting accompanied by extensive deformation. Shoreline exposure from the small unnamed island in Tracy Arm, near Sawyer Glacier.

Rocks in the central pluton-gneiss belt and eastern edge of the western metamorphic belt were metamorphosed at pressures of 8,000 to 10,000 times atmospheric pressure and temperatures approaching 1,292°F (700°C). These conditions are representative of about 18 miles (30 km) below the surface or the middle to lower crust. Therefore, an amazing 15 to 18 miles (25 to 30 km) of rock must have been eroded from above the current land surface. Chronological studies indicate that the rocks cooled at rates of over 90°F (50°C) per million years.[41] This suggests rapid exhumation rates of about 1/16 inch (1.5 mm) per year shortly after intrusion of the tonalite sill about 60 million years ago.

The Tracy Arm and Windham Bay area form the southern end of the Juneau gold belt. Numerous placer and hard rock gold mines form a linear belt within rocks of the western metamorphic belt.[42]

The abandoned mines at Windham Bay, Snettisham, and Sumdum were some of the earliest (1870) deposits found and exploited in the gold belt. All of these mines followed lode-quartz veins into the mountainsides; however, much of the recent mineral exploration has examined massive sulfide layers that contain appreciable amounts of gold and silver. Sulfide deposits have been explored along the south shore of Endicott Arm, near the top of Mt. Sumdum, just south of the entrance to Tracy Arm, and on the ridge north of the entrance to Tracy Arm known as Sweetheart Ridge.

My geological career began in mineral exploration. During the summer of 1979, I worked for Mid America Pipeline Company. We carried out gold and silver exploration throughout Southeast Alaska. One of the company's prospects was Sweetheart Ridge: a gold-bearing sulfide layer located 3,000 feet (900 m) above Tracy Arm (Figure 29). For four summers, we carried out mapping, geochemical sampling, and diamond core drilling that successfully delineated a small gold-bearing ore body. These summers were an incredible geological and wilderness experience. We camped above tree line with a great view of the fjord and the alpine ridges above it. The small gold, silver, copper, lead, and zinc ore body lies along the alpine ridge north of the entrance to Tracy Arm (Figure 29). This beautiful locally knife-sharp ridge, which we affectionately called Sweetheart Ridge for the narrow lake on the north side, forms the boundary between the Tracy Arm–Fords Terror Wilderness Area to the south and the less-well-protected Tongass Forest land to the north. Extraction of the ore would require blasting large trenches above tree line, construction of a road down to Gilbert Bay, and a dock on the bay. It was certain that this would be a major disruption of the ecosystem that was home to numerous bears, old growth trees, and salmon among many other species. My position as geologist during the summer of 1982 provided me with a personal dilemma. On the one hand, I was keen on development of the resource that I had worked on because it would be positive for the company and my career. But on the other hand, I grew to love the land, plants, and animals more than any gold. It was with some relief that I was able to recommend cancellation of the project to the company management due to low tonnage and complicated geometry of the ore body.

During the heyday of gold mining, around the turn of the century, booming towns thrived at Snettisham, Sumdum, and Windham (Figure 29). All of these towns are long abandoned and little remains of the log structures that were present. During the 1930s entrepreneurs established a fox farm on Harbor Island. Small islands throughout Southeast Alaska were favored as sites for these farms because there the foxes could be easily controlled and the only predators (bears) were either absent or could be eliminated. Today, much of the area is protected as wilderness and only the areas around Snettisham and Windham are outside the Tracy Arm–Fords Terror Wilderness Area.

Frederick Sound

Frederick Sound is a wide body of water that extends east from Chatham Strait to the mainland and then continues southeast between Kupreanof and Mitkof islands and the mainland (Figure 1). Northern Kuiu and Kupreanof islands border the sound to the south and Admiralty Island forms the northern shore. Most of the shoreline is sparsely inhabited, but the small town of Petersburg lies at the southeast end of Mitkof Island. Frederick Sound is well known for the summer populations of humpback whales and Steller's sea lions that congregate around the northwest end of the sound where food is abundant.

The southern tip of Admiralty Island is dominated by massive volcanic flows, known as the Admiralty Island Volcanics, which poured out onto the land during early Tertiary time. The andesite and basalt flows are estimated to be around 10,000 feet (3,000 m) thick. These impressive volcanic rocks can be readily observed from Hood and Pybus bays on southern Admiralty Island. Underlying the volcanic rocks are slightly metamorphosed Paleozoic sediments of the Alexander terrane. Northern Kuiu and Kupreanof islands are also formed from Silurian to Permian sedimentary rocks of the Alexander terrane. The low cliffs that form the shorelines of these two islands are horizontal layers of gray limestone. On the northeast shore of Kuiu Island the ocean has carved this limestone into sea stacks and arches that contain abundant evidence of shallow marine life during the late Paleozoic. Marine fossils include snails, corals, clams, and other somewhat similar shellfish called *brachiopods*. The clams are not dissimilar to those that are common today. The brachiopods also have two-piece shells, but differ in that their plane of symmetry crosses through the shells. The limestone cliffs on the east of Saginaw Bay (Figures 22 and 32) are still adorned by symbols placed there by Native Americans. Further east, mudstone and greywacke of the Gravina belt line the northern shore of Kupreanof Island.

The western end of Frederick Sound meets the linear trend of Chatham Strait where rocks of the Alexander terrane terminate along the Chatham Strait fault. The eastern end of the sound meets the high peaks of the Coast Mountains underlain by igneous rocks of the Coast Plutonic Complex.

Petersburg and Wrangell

Petersburg is a small fishing town built on the northeast corner of Mitkof Island. This busy and pleasant town has a wonderful view of the Coast Mountains and the upper parts of the Stikine Icefield east of town across Frederick Sound. Proximity to

FIGURE 32. Sea arch carved into Permian limestone cliffs by wave action along the coast of Kuiu Island.

the Stikine Icefield places Petersburg within 30 miles (48 km) of tidewater glaciers located up narrow fjords at Thomas Bay and LeConte Bay (Figures 1 and 17). The northern and western sides of town are bordered by a shallow channel of water known as the Wrangell Narrows, which separates Mitkof and Kuiu islands. This channel allows small to medium-sized ships to enter the town's harbor from the south through a convoluted path of navigation lights sometimes called "the Christmas trees." Petersburg has yet to change much with growth of tourism, in part because the inhabitants have not chosen to attract large ships into their town.

West of Petersburg across the Wrangell Narrows lies Kupreanof Island, which is underlain by sedimentary rocks of the Alexander terrane on the northeast and Tertiary volcanic rocks on the southwest.

Garnet Samarium and Neodymium Geochronology

Samarium, like uranium, has radioactive isotopes that are unstable and decay to form new elements as a result of changes that occur in the nucleus of the atom. For example, the samarium isotope with an atomic weight of 147 decays to neodymium with an atomic weight of 143. The rate of atomic decay is often stated as the time required for one half of the atoms to decay, or half-life. In the case of samarium 147, the half-life is 106 billion years, making samarium useful for dating very old events because even after millions of years measurable samarium will remain and a precise age can be calculated.

Both samarium and neodymium are incorporated into garnet during crystal growth. As a result, the ratio of these two elements cannot be used to directly calculate an age. Most commonly, the age is calculated by determining the isotopic composition of more than one mineral to find a line whose slope depends on age.

Core and rim samples from single garnet crystals can be dated separately. Assuming that the crystals grow concentrically outward from the center, then ages determined for the core and rim of crystals provide estimates for each stage of growth. In this case, when the duration of growth and crystal size is considered, the rate of growth can be determined.

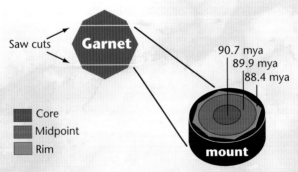

DIAGRAM OF AGES for garnet core, garnet midpoint, garnet rim, and rock samples from Garnet Ledge, Alaska. The isotope ages indicate that garnet grew between 90.7 and 88.4 million years ago. From Stowell et al. (2001).

Southeast of Petersburg, the Stikine River has eroded a broad valley through the Coast Mountains and deposited a wide tidal delta crossed by numerous channels braiding through the shallow water. Small islands largely underlain by intrusive diorite bodies that are resistant to erosion dot this broad delta and Sumner Strait to the west.

Petersburg lies along the western edge of low-grade metamorphic rocks of the western metamorphic belt. These Gravina belt rocks are dominantly clastic sediments.[43] However, higher peaks on Mitkof Island and the Lindenburg Peninsula of nearby Kupreanof Island are underlain by diorite intrusive rocks that are far more resistant to weathering and erosion than the metamorphosed sediments. These plutons intruded the Gravina belt rocks about 100 million years ago. Looking east across Frederick Sound from Petersburg, older metamorphosed sedimentary and volcanic rocks of the Taku terrane form the lower mountains in the foreground, and 50-to-60-million-year-old granite, tonalite, and high-temperature metamorphic gneiss of the central pluton-gneiss complex underlie the highest peaks, including Devil's Thumb, of the Coast Mountains in the background (Figure 33).

FIGURE 33. Photograph from Petersburg looking east toward the high peaks of the Coast Mountains, which are underlain by Paleocene tonalite of the Coast Plutonic Complex sill and the central pluton-gneiss belt.

LeConte Bay is a short but beautiful fjord that extends eastward from the southern part of Frederick Sound into the high peaks of the Coast Mountains (Figure 1). The entrance to the fjord is nearly blocked by an end moraine that accumulates numerous icebergs as the tide moves in and out. After a sinuous journey up the fjord past the moraine and numerous icebergs, visitors are treated to a view of LeConte Glacier, the southernmost tidewater glacier in North America and one of the most active glaciers in the Coast Mountains. Recent observations indicate that the glacier has retreated more than half a mile since 1994 and has flow rates that vary from 25 to 95 feet (8 to 29 m) per day.[44] Rapid retreat combined with this high rate of flow during the past ten years has produced so many icebergs that boats are often unable to enter the fjord.

The small town of Wrangell lies southwest of the Stikine delta on the northern tip of Wrangell Island (Figures 1 and 34). Wrangell, like Petersburg, has changed little with growth of tourism and retains much of its old character. Wrangell is also built on rocks of the Gravina belt. This part of the belt was metamorphosed at low temperatures (less than about 572°F or 300°C). The diorite plutons on the islands and mainland around and west of the Stikine delta have narrow belts of metamorphic minerals around them that grew as a result of heat from the plutons. These belts of minerals are known as contact metamorphic *aureoles*. Several interesting and important minerals are found in these aureoles. Several of the aureoles contain one or more of the three aluminum silicate minerals: *kyanite*, *andalusite*, and sillimanite. Some of these aureoles

▶ **FIGURE 34.** Simplified geology of the Petersburg and Wrangell areas. Intrusive igneous rocks are shown in red, Gravina belt rocks are shown in tan, western metamorphic belt rocks are shown in blue, and Coast Plutonic Complex rocks are shown in brown. Lines show the extent of regional metamorphic mineral zones. Geology modified from Brew and others (1985) and Douglass and Brew (1985).

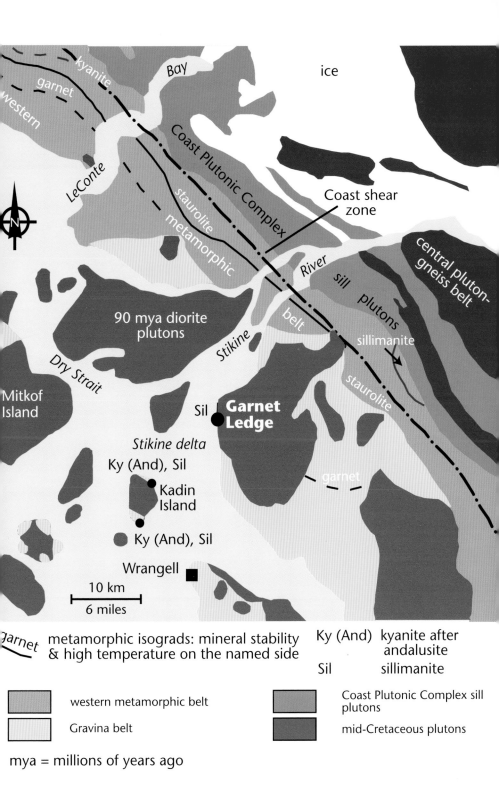

ice

Bay

kyanite

garnet

western

LeConte

Coast Plutonic Complex

staurolite

metamorphic

Coast shear
zone

River

sill

central pluton-
gneiss belt

N

90 mya diorite
plutons

Stikine

belt

plutons

sillimanite

Dry Strait

staurolite

Mitkof
Island

Sil

Garnet
Ledge

Stikine delta

Ky (And), Sil

garnet

Kadin
Island

Ky (And), Sil

Wrangell

10 km

6 miles

garnet   metamorphic isograds: mineral stability   Ky (And)   kyanite after
          & high temperature on the named side   andalusite
                                                      Sil   sillimanite

western metamorphic belt   Coast Plutonic Complex sill
                                            plutons
Gravina belt   mid-Cretaceous plutons

mya = millions of years ago

contain kyanite replacing andalusite. Because kyanite requires higher pressure or lower temperature than andalusite to grow and minerals tend to grow during increasing temperature, the kyanite has been used to infer an increase in pressure. The most likely explanation for this is that northwest-trending faults, including the Coast shear zone (Figure 34), thrust rocks to the southwest, increasing the weight of rocks above the contact aureoles and thickening the crust. Garnet is another abundant mineral in the contact aureoles. One of the aureoles, known as Garnet Ledge, located along the southern shore of the Stikine River (Figure 34), contains so much garnet that it was once mined for it.

Garnet Ledge is a garnet-rich body of contact metamorphic rocks (Figures 34 and 35) on the north side of Garnet Mountain. The rocks are so rich in garnet that they were mined for this mineral around 1915. Intriguingly, the mine was owned and operated by the first all-female corporation in the U.S. and perhaps in the world (Figure 35b). Currently, the garnet-bearing rocks are owned by the youth of Wrangell, who extract small quantities of the mineral and sell them to tourists. The garnet, which is very attractive and hard, was used for making jewelry and abrasives (for example, sandpaper).

Geologists often study garnet because it is relatively common, its chemistry depends on the temperature and pressure at which it grows, and the chemistry of the mineral tends to be preserved even during metamorphism. In the last ten years, a few geologists have used isotopes of samarium and neodymium to date the growth of individual garnet crystals. David Taylor (a former graduate student at the University of Alabama) and I studied the garnet at Garnet Ledge and the aureoles on Kadin and Mitkof islands.[45] Our geochemical and isotope data indicate that garnet grew at temperatures of about 1,200°F (650°C) and individual crystals grew at about the same time as intrusion of the pluton nearby. This is no great surprise, but more interestingly, ages calculated for several segments (core or center and rim or edge) of single crystals indicate that the

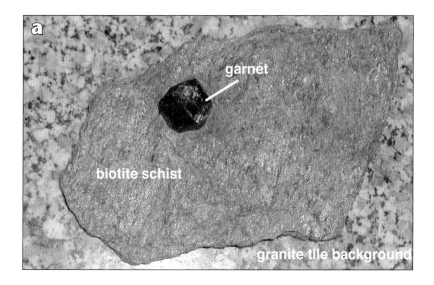

FIGURE 35. Large (up to 1.5 inch or 3.7 cm) garnet crystals from Garnet Ledge are dominantly almandine—an iron-rich garnet that is common in metamorphic rocks. (a) Approximately 1-inch (25-mm) diameter garnet crystal in *biotite* schist. (b) Anna E. Durkee, one of the early garnet miners, holding crystals extracted along Garnet Creek, probably about 1906. (Courtesy Wrangell Museum.)

crystal grew over a period of one to two million years. The crystals that we studied are about one inch (2.5 cm) in diameter. Therefore, these garnet crystals must have grown at a rate of about ¼ to ½ inch (6 to 12 mm) per million years.

The Stikine River carries vast amounts of sand and mud west across the Coast Mountains from central British Columbia. Where this large river flows into the ocean northeast of Wrangell (Figure 34), this is deposited in a fan-shaped array of shallow channels between the small islands north of the town. The abundant fine sediment created by glaciers in the Coast Mountains to the east has largely filled the former channels, resulting in the treacherous passage known as Dry Strait (Figure 34). Most mariners avoid these waters because they are only passable in a small boat at high tide.

Clarence Strait and Prince of Wales Island

Clarence Strait is a long body of water that separates Prince of Wales Island on the west from smaller islands and the mainland on the east (Figure 1). Although there are no significant towns along the shores, the strait serves as a major artery of the Inside Passage with numerous ferries and commercial and private boats plying the water. Large parts of Prince of Wales Island have been logged and the rectangular patches of cleared forest can be seen from Clarence Strait. As a result of the logging, Prince of Wales Island has an extensive network of roads and some of these are accessible to the public.

► **FIGURE 36.** Simplified geology of the Ketchikan area. Intrusive igneous rocks are shown in medium red, Gravina belt rocks are shown in tan, western metamorphic belt rocks are shown in blue, and central pluton-gneiss belt are shown in dark red. Modified from Crawford and others (2000).

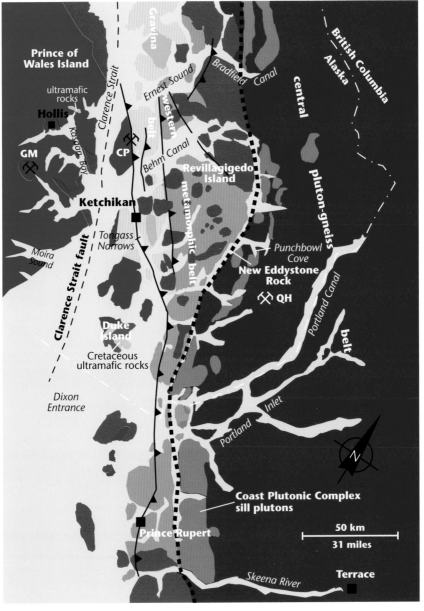

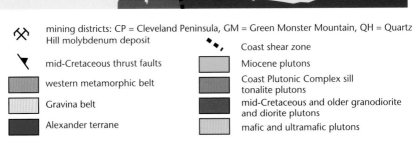

mining districts: CP = Cleveland Peninsula, GM = Green Monster Mountain, QH = Quartz Hill molybdenum deposit

Coast shear zone

mid-Cretaceous thrust faults

Miocene plutons

western metamorphic belt

Coast Plutonic Complex sill tonalite plutons

Gravina belt

mid-Cretaceous and older granodiorite and diorite plutons

Alexander terrane

mafic and ultramafic plutons

The eastern shore of Clarence Strait is formed by Paleozoic Alexander terrane rocks south of Ernest Sound, and Mesozoic Gravina belt rocks to the north (Figures 17 and 36). The western shore of the strait is entirely underlain by Alexander terrane rocks and a variety of Paleozoic to Mesozoic plutons (Figure 36). The metamorphosed basalt and greywacke near the town of Hollis is the oldest section of the Alexander terrane, and although no precise age is known, it must be pre-Ordovician. In central Prince of Wales Island, the metamorphic rocks are intruded by granite that has produced unusual occurrences of radioactive minerals at Bokan Mountain and beautiful crystals of garnet, calcite, and quartz, and world-class specimens of a green mineral called *epidote* at Green Monster Mountain (Figures 36 and 37).

Ketchikan and Misty Fiords

Ketchikan is situated in the southernmost part of Southeast Alaska and is often considered a gateway city for travelers traveling north from the lower forty-eight states on the Inside Passage (Figure 1). The city lies along the eastern shore of Tongass Narrows between Revillagigedo and Gravina islands (Figure 36) and is about 20 miles (32 km) west of Misty Fiords National Monument. Ketchikan has a long history as a fishing and logging town, but with closure of the local pulp mill and growth of tourism, the town has changed dramatically. For example, souvenir shops have largely replaced the bars and somewhat seedy boardinghouses that once dominated the downtown area. Similar to Petersburg and Juneau, Ketchikan is underlain by rocks of the western metamorphic belt and is situated along the eastern boundary of the Gravina belt. However, unlike its northern counterparts, Ketchikan is now far south of the remaining tidewater glaciers in Southeast Alaska.

FIGURE 37. Large crystals of dark green epidote, clear quartz, and white calcite from Green Monster Mountain, Prince of Wales Island. These unusually large crystals formed in open cavities when a nearby granitic magma caused metamorphism of the surrounding quartz-rich limestone. Photo is from the display case in the Juneau airport.

The Ketchikan area has a long mining history with deposits of copper, molybdenum, uranium, and iron found nearby. Mines and known mineral deposits stretch from the Coast Mountains north and east of town to Duke and Prince of Wales islands. The Cleveland Peninsula was extensively explored for gold during the late nineteenth and early twentieth centuries. Quartz Hill east of Behm Canal (Figure 36) is one of largest known molybdenum deposits in the world. The deposit has never been mined owing to environmental concerns and low-cost production of molybdenum from other places. Compilations of these locations and brief descriptions of the mineral deposits can be found through the Juneau Mineral Information Center.

Perhaps the most striking thing one notices on arrival in Ketchikan is the predominance of green rocks. This is particularly obvious in the gravel along the airport runway on Gravina Island or south of the cruise ship terminal on Revillagigedo Island. At these locations, the recently crushed metamorphosed basalt truly deserves the name of greenstone. The bright coloration is largely derived from chlorite and epidote that grew during low- to medium-temperature metamorphism. Gravina Island is the first place where rocks of the sediments and volcanic rocks of the Gravina belt were described. Therefore, this is called the "type locality" and provides the name. Slightly metamorphosed volcanic rocks that range from basalt to rhyolite, which formed in a late Cretaceous volcanic arc, underlie most of Gravina Island. Revillagigedo Island is underlain by medium-temperature metamorphic rocks of the Taku terrane, which are intruded by Cretaceous granitic rocks and Miocene *gabbro*. Juxtaposition of the Gravina and Taku terranes occurred along a northwest-oriented reverse fault that transported the Taku terrane southwest over the Gravina belt (Figure 36), analogous to the movement along the Gastineau Channel fault in Juneau. Not only is the fault along Tongass Narrows similar to the Gastineau

Channel fault in the juxtaposition of lithologies, but also in local-ized gold deposits formed by fluid flow along the fault.

South of Ketchikan, Duke Island lies about 10 miles (16 km) from the mainland (Figure 36). Although not impressive from a distance, this island is well known for its unusual iron- and magnesium-rich igneous rocks. These Cretaceous rocks, known as Alaska-type zoned ultramafic plutons, were intruded along a linear belt from Duke Island up to Haines. Rocks in the plutons are mostly *dunite*, com-posed of the yellow-green mineral olivine, to *peridotite*, composed of *pyroxene* and olivine. These ultramafic rocks are thought to have formed when the iron- and magnesium-rich minerals crystallized along the feeders to now-eroded volcanoes. In addition to the sili-cate minerals, many of these plutons contain significant amounts of the iron-oxide mineral *magnetite*. The strong magnetic field pro-duced by the magnetite here is strong enough to cause compass bearings to be 90 degrees from true north!

Misty Fiords, east of Ketchikan (Figure 1), is underlain by meta-morphic rocks of the Taku terrane and Tertiary tonalite plutons that intruded along or near the border between the western meta-morphic belt and central pluton-gneiss belt to the east (Figure 36). The entrance to Misty Fiords is framed by spectacular cliffs, many of which are formed by the Coast Plutonic Complex sill (Figure 38). The same linear belt of erosion-resistant tonalite intrusive rocks extend a remarkable 400 miles (640 km) from south of Ketchikan to Skagway. Although tidewater glaciers have long been absent, their retreat has left a remarkable array of glacially carved topog-raphy that includes innumerable U-shaped valleys and arêtes. The most commonly traveled of the fjords, Behm Canal, provides a long and beautiful loop through the heart of the region and the center of the Coast Plutonic Complex (Figure 36). Visitors travel-ing into the southern entrance of Behm Canal are greeted by the steep 234-foot (71-m) high tower of New Eddystone Rock (Figure

FIGURE 38. Cliffs formed by one of the Coast Plutonic Complex sill tonalite plutons near the boundary of Misty Fiords National Monument, Punchbowl Cove.

FIGURE 39. The 234-foot (71-m) high tower of basalt known as New Eddystone Rock, Behm Canal in Misty Fiords National Monument.

39), the erosion-resistant core of a conduit that once transported magma toward the surface. Rocks in the New Eddystone area are basaltic[46] and are part of an extensive swarm of Tertiary *dikes* that intruded the Coast Mountains after the majority of exhumation had ended. These dikes may indicate spreading or extension of the crust after cessation of mountain building.

Notes

1. Barron, "Paleoclimatology."
2. Ibid.
3. Ibid.
4. Pang and Yau, "Ancient Observations Link Changes in Sun's Brightness and Earth's Climate."
5. EPICA Community Members, "Eight Glacial Cycles from an Antarctic Ice Core."
6. Ruddiman, *Earth's Climate: Past and Future.*
7 Haeberli and others, *World Glacial Inventory.*
8. Arendt and others, "Rapid Wastage of Alaska Glaciers and Their Contribution to Rising Sea Level."
9. Bowditch, *The American Practical Navigator.*
10. See note 8.
11. Schoenfeld and Thompson, "Caves on Prince of Wales Island."
12. Romm, "New Forerunner for Continental Drift."
13. Kearey and Vine, *Global Tectonics.*
14. Gore, "Cascadia: Living on Fire."
15. Stowell and McClelland, "Tectonics of the Coast Mountains."
16. Eberlein and Churkin, "Paleozoic Stratigraphy in the Northwest Coastal Area of Prince of Wales Island, Southeastern Alaska."
17. Hollister and Andronicos, "A Candidate for the Baja British Columbia Fault System in the Coast Plutonic Complex."
18. Morozov et al., "Generation of New Continental Crust and Terrane Accretion in Southeastern Alaska and Western British Columbia: Constraints from P- and S-wave Wide-angle Seismic Data (ACCRETE)."

19. Spotila et al., "Long-term Glacial Erosion of Active Mountain Belts: Example of the Chugach–St. Elias Range, Alaska."

20. Engebretson et al., "Relative Motions Between Oceanic and Continental Plates in the Pacific Basin."

21. Berg et al., "Gravina Nutzotin Belt—Tectonic Significance of an Upper Mesozoic Sedimentary and Volcanic Sequence in Southern and Southeastern Alaska."

22. Cohen and Lundberg, "Detrital Record of the Gravina Arc, Southeastern Alaska: Petrology and Provenance of Seymour Canal Formation Sandstones."

23. See notes 15 and 21.

24. Larsen et al., "Rapid Uplift of Southern Alaska Caused by Recent Ice Loss."

25. Dauenhauer and Dauenhauer, *Haa Shuká, Our Ancestors: Tlingit Oral Narratives.*

26. Brew, Horner, and Barnes, "Bedrock-geologic and Geophysical Research in Glacier Bay National Park and Preserve: Unique Opportunities of Local-to-Global Significance."

27. Fletcher and Freymueller, "New Constraints on the Motion of the Fairweather Fault, Alaska, from GPS Observations."

28. O'Sullivan, Plafker, and Murphy, "Apatite Fission-track Thermotectonic History of Crystalline Rocks in the Northern St. Elias Mountains, Alaska."

29. Gehrels, "Reconnaissance Geology and U-Pb Geochronology of the Western Flank of the Coast Mountains between Juneau and Skagway, Southeastern Alaska."

30. Begét and Motyka, "New Dates on Late Pleistocene Dacitic Tephra from the Mount Edgecumbe Volcanic Field, Southeastern Alaska."

31. Hudson, Plafker, and Dixon, "Horizontal Offset History of the Chatham Strait fault."

32. Nolan et al., "Ice-thickness Measurements of Taku Glacier, Alaska, and Their Relevance to its Dynamics."

33. Stone and Stone, *Hard Rock Gold: The Story of the Great Mines that were the Heartbeat of Juneau.*

34. Stowell, Menard, and Ridgway, "Ca-metasomatism and Chemical Zonation of Garnet in Contact Metamorphic Aureoles, Juneau Gold Belt, Southeastern Alaska."

35. Goldfarb et al., "Origin of Lode-gold Deposits of the Juneau Gold Belt, Southeastern Alaska."

36. Stowell, "Sphalerite Geobarometry in Metamorphic Rocks and the Tectonic History of the Coast Ranges Near Holkham Bay, Southeastern Alaska"; Stowell, "Silicate and Sulphide Thermobarometry of Low- to Medium-grade Metamorphic Rocks from Holkham Bay, South-east Alaska"; Wood et al., "^{40}Ar/^{39}Ar Constraints on the Emplacement, Uplift, and Cooling of the Coast Plutonic Complex Sill, Southeastern Alaska."

37. Gehrels et al., "Ancient Continental Margin Assemblage in the Northern Coast Mountains, Southeast Alaska and Northwest Canada."

38. See note 29.

39. Stowell, Green, and Hooper, "Geochemistry and Tectonic Setting of Basaltic Volcanism, Northern Coast Mountains, Southeastern Alaska."

40. Brew and Grybeck, "Geology of the Tracy Arm–Fords Terror Wilderness Study Area and Vicinity, Alaska."

41. Wood et al., "^{40}Ar/^{39}Ar Constraints on the Emplacement, Uplift, and Cooling of the Coast Plutonic Complex Sill, Southeastern Alaska."

42. Brew et al., "Preliminary Reconnaissance Geologic Map of the Petersburg and Parts of the Port Alexander and Sumdum 1:250,000 Quadrangles, Southeastern Alaska."

43. Cohen and Lundberg, "Detrital Record of the Gravina arc, Southeastern Alaska: Petrology and Provenance of Seymour Canal Formation Sandstones."

44. White et al., "1999 Results of Student Field Survey at Leconte Tidewater Glacier, Southeastern Alaska."

45. Taylor, "Samarium-neodymium Garnet Geochronology and Silicate Mineral Thermobarometry of Late Cretaceous Metamorphism, Garnet Ledge, Southeast Alaska"; Stowell et al., "Contact Metamorphic P-T-t Paths from Sm-Nd Garnet Ages, Phase Equilibria Modeling, and Thermobarometry: Garnet Ledge, Southeastern Alaska."

46. Ouderkirk, "Tertiary Mafic Dikes in the Northern Coast Mountains, British Columbia."

Glossary of Geological Terms

accretion: Addition of material to a continent by convergent and transform motion of island arcs and continental fragments.

accretionary prism: A wedge-shaped mass of deformed sediment above a subduction zone. The Chugach terrane on the west side of the Alexander Archipelago is an example of an accretionary prism.

alluvial fan: Cone-shaped deposits of gravel and sand deposited by a stream at any location where the stream velocity decreases due to a decrease in gradient.

altimetry: The subject of measuring height above sea level or another reference surface. Laser altimetry is being used to precisely measure changes in elevation and shapes of surfaces.

amphibole: A group of elongate minerals with highly variable chemistry that form at low to high temperature in igneous and metamorphic rocks and are comprised of oxygen, silicon, aluminum, iron, magnesium, manganese, calcium, potassium, sodium, and water.

andalusite: A low-pressure mineral composed of aluminum, silicon, and oxygen (Al_2SiO_5). Andalusite is most commonly formed by metamorphism of mudstone.

andesite: A volcanic igneous rock with intermediate silica and low iron and magnesium content. These rocks are commonly erupted from the tall steep volcanoes above subduction zones.

113

arête: Narrow ridges left remaining between glacially carved U-shaped valleys. These steep-sided ridges often provide beautiful locations for alpine walks and views.

aureole: Metamorphic rocks formed as a result of heat from a cooling igneous body. Rocks in the aureole are often called contact metamorphic rocks.

basalt: A volcanic igneous rock with low silica and high iron and magnesium content. These rocks are commonly erupted from broad but large volcanoes like those in Hawaii or from long fissures or cracks.

bergy bit: Small pieces of floating glacial ice that are smaller than an iceberg and larger than a growler. Bergy bits extend between 3.3 and 16.7 feet (1 to 5 m) above sea level and are generally less than 32.8 feet (10 m) across.

biotite: A dark platy mineral that contains potassium, iron, magnesium, aluminum, silicon, and oxygen. Biotite is common in igneous rocks and is the primary mineral in many metamorphic rocks.

brachiopod: A type of shellfish with two shells. The shells have mirror symmetry across a plane that cuts through the center of the shells. Although some brachiopods survive today, they are known for their abundance during the Paleozoic and Mesozoic.

calcite: A common mineral composed of calcium, carbon, and oxygen ($CaCO_3$). Calcite is the defining constituent of limestone and marble. Large areas of limestone underlie Prince of Wales Island in Southeast Alaska.

calving: The formation of icebergs by collapse of glacial ice into water.

chemical sediment: Sediment formed by organic or inorganic precipitation of ions from solution. Hardening of this sediment forms a chemical sedimentary rock.

chlorite: A group of platy green minerals with variable chemistry that forms at low to medium temperature. The most common forms are found in metamorphic rocks and are comprised of oxygen, silicon, aluminum, iron, magnesium, manganese, and water.

cirque: A steep-walled bowl-shaped depression carved by glacial ice.

clastic: Rock and mineral fragments that are transported as solid pieces and then deposited to make sediment. Hardening of this sediment forms a clastic sedimentary rock.

cordillera: A term used to describe a group of subparallel mountain ranges.

crevasse: A fracture in the upper brittle part of a glacier or icefield.

crust: Outermost layer of the earth which is divided into rigid plates. There are two types of crust: oceanic, composed of silica-poor igneous rocks (for example, basalt), and continental, composed of silica-rich rocks (for example, granodiorite).

deposition: Accumulation of clastic or chemical sediment in water or air.

dike: Tabular igneous rock bodies that crystallize in a fracture within solid rocks.

diorite: Igneous rock with intermediate silica content that typically contains plagioclase, quartz, biotite, and hornblende.

dunite: Coarse-grained igneous rock dominantly made of the pea green mineral olivine. This is an ultramafic rock because it has a very low silica (SiO_2) content and high magnesium and iron content.

epidote: A medium to dark green calcium and iron silicate mineral. Beautiful specimens have been found in metamorphic rocks produced by intrusion of granite at Green Monster Mountain on Prince of Wales Island.

erosion: Removal of material from rocks at the earth's surface. This process involves mechanical breakup or chemical breakdown of the rocks.

erratics: Rock fragments carried by flowing ice and deposited far from the rocks from which they are derived.

exhumation: The combination of erosion and uplift that exposes formerly buried rocks at the surface.

exotic terrane: Pieces of crust that have a distinct history which is different than that of the adjacent rocks. In some cases, fossils and paleomagnetism suggest that these terranes may have moved far from their current location. Numerous terranes are present in southern and Southeast Alaska.

fault: Any fracture in rock along which displacement has occurred. Faults may have mostly up or down (dip slip) or mostly horizontal (strike slip) displacement.

feldspar: A group of blocky light-colored minerals with variable chemistry that form at low to high temperature. The most common forms are comprised of oxygen, silicon, aluminum, and calcium. This group is subdivided into plagioclase and alkali feldspars.

felsic: Silica (SiO_2)-rich igneous rocks (for example, rhyolite or granite).

ferromagnesium: Word used to describe minerals and rocks that contain a lot of iron and magnesium.

fjord: A steep U-shaped valley carved by a glacier that has been flooded by ocean water.

folding: Process of bending planar features in a rock to produce a curved surface.

foliation: A planar fabric or layering in rocks caused by variation in mineralogy or mineral alignment. Foliation is typical of regional metamorphic rocks.

foraminifera: Small marine or freshwater organisms that generally form shells made of calcite. These creatures have inhabited seas and lakes since the Cambrian period of earth history.

gabbro: An iron- and magnesium-rich igneous rock with large crystals formed by slow-cooling of magma with a basaltic composition.

garnet: An equant mineral with variable chemistry that forms at medium to high temperature. The most common forms are found in metamorphic rocks and are comprised of oxygen, silicon, aluminum, iron, magnesium, manganese, and calcium. I have used this important mineral widely in my studies to understand crustal structure and processes.

geochronology: Science of determining the age of earth events and materials. Most modern age determinations are based on radioactive decay of elements and the resulting isotopic ratios.

geomorphology: Study of the earth's surface, including the shape of the surface and the processes associated with its formation.

geophysics: Study of the earth by quantitative methods, including study of earthquake energy, magnetism, gravity, and electrical conductivity. These techniques provide important information about the earth's interior.

glacial striations: Grooves cut in the rock by flowing glacial ice; generally caused by boulders dragged across the surface by the flowing ice.

glaciation: A term used to describe all glacial processes. This word is also used to describe periods of time in which large-scale ice sheets covered large areas of continents.

glacier: Generally defined as ice that shows evidence of past or present movement. Alpine or valley glaciers are fed by snow-covered mountains and flow down valleys. Continental glaciers or ice sheets are unconfined and may cover an entire continent.

gneiss: A coarsely layered metamorphic rock that forms at moderate to high metamorphic temperatures (500 to 700°C).

granite: An intrusive or plutonic igneous rock found in mountain belts. Light-colored plagioclase and quartz, pink potassium feldspar, and dark *ferromagnesium* minerals give this rock a tricolor appearance.

granodiorite: An intrusive or plutonic igneous rock commonly found in mountain belts around the world. Light-colored plagioclase and quartz and dark ferromagnesium minerals commonly give this rock a salt-and-pepper appearance.

greenhouse effect: An atmospheric process that causes retention of solar energy and results in warming of the earth. Various gases such as carbon dioxide, water, and methane cause this phenomenon which occurs naturally and could be increased by humans' output of gas.

greenstone: A massive green metamorphic rock generally resulting from low temperature (572 to 842°F or 300 to 450°C) metamorphism of a volcanic rock (generally basalt).

greywacke: A poorly sorted coarse-grained immature sandstone that is largely composed of quartz and feldspar, plus or minus ferromagnesian silicate minerals and rock fragments. Greywacke forms when dense sediment-laden bottom currents deposit materials in the deep ocean adjacent to a continent. Greywacke is the Anglicized form of the German word *grauwacke* and means gray stone.

growler: A small piece of floating glacial ice that is smaller than a bergy bit. Growlers extend less than 3.3 feet (1 m) above sea level and are generally less than 20 feet (6 m) across.

hanging valley: A U-shaped valley that joins another valley, but whose mouth is far above the valley it joins.

horn: A sharp-pointed mountain that is carved by glaciers flowing away from a central high area (for example, the Matterhorn).

hornblende: A dark green, elongate, medium-to-high-temperature silicate mineral. The mineral may be metamorphic or igneous in origin and is composed of sodium, calcium, iron, magnesium, aluminum, silicon, oxygen, and hydroxyl.

ice age: A period of earth history in which global ice volumes were much greater than present. Average global temperatures were probably only a few degrees less than today.

iceberg: Large pieces of glacial ice that calve or break off from a glacier into water. Icebergs extend more than 16.7 feet (5 m) above water and are generally more than 32.8 feet (10 m) across.

icefield: Large, relatively flat areas of ice that typically feed numerous glaciers that flow away from the icefield.

igneous rock: Rocks that form from crystallization of a liquid rock or magma. Almost all igneous rocks are dominantly comprised of silica and aluminum with lesser amounts of iron, magnesium, potassium, and sodium oxides.

intermediate rock: An igneous rock with an intermediate silica content (for example, andesite or diorite).

intrusion: The processes of emplacing magma into pre-existing rocks and forming plutonic igneous rocks. These igneous rocks typically cool slowly and have large crystals. Much of the Coast Mountains are underlain by solidified magma that intruded between 80 and 50 million years ago, known as the Coast Plutonic Complex.

isostasy: The equilibrium or floating of crust on the earth's mantle which flows slowly like a plastic. Thicker or less dense crust floats higher, resulting in mountains.

isotopes: Atoms with the same number of protons (one element) that have different numbers of neutrons.

kyanite: A high-pressure mineral composed of aluminum silicon and oxygen (Al_2SiO_5). Kyanite is most commonly formed by metamorphism of mudstone.

limestone: Sedimentary rock dominantly composed of calcite. Most limestone is deposited by biological activity.

lineament: Linear features on the land surface formed due to erosion of rocks that are comprised by adjacent areas of harder and softer rocks. Many lineaments are formed due to erosion of broken-up rocks formed by fault movement. The Coast Range megalineament is a prominent topographic low formed by erosion of broken-up rocks in faults along the west flank of the Coast Mountains.

lithosphere: (a) The solid part of the earth. (b) The outer rigid and strong material near the earth's surface that lies above more plastic, weaker material underneath.

lode gold: Gold found in vein and disseminated deposits in solid rocks.

mafic rock: Silica-poor igneous rocks (for example, basalt or gabbro).

magma: A hot liquid rock that may reach the surface and form volcanic rock or solidify beneath the surface and form plutonic rock.

magmatism: All of the processes involving magma formation, movement, and crystallization.

magnetite: An oxide mineral with unusually strong magnetic properties composed of iron and oxygen (Fe_3O_4).

mantle: Layer of the earth beneath the crust. This layer of silicate rock extends from the crust to the liquid core.

marble: A metamorphic rock composed of carbonate minerals (for example, calcite). This rock generally forms when limestone recrystallizes due to increased temperature.

mélange: Rocks that are characterized by a lack of continuous internal layers and inclusions of rock fragments and blocks of all sizes in a matrix. These rocks may form by sedimentary processes or tectonic disruption involving faults. Derived from the French word meaning mixture.

metamorphic rock: Any rock that forms within the earth from solid-state recrystallization of pre-existing rock at high temperature and/or pressure.

metamorphism: Solid state change of rocks beneath the earth's surface due to changes in pressure, temperature, and/or chemical environment.

mica: A group of silicate minerals that have a tabular form and a layered internal structure that imparts the ability to break along parallel closely spaced planes. Biotite and muscovite are two of the most common mica minerals.

migmatite: A composite rock composed of mixed metamorphic and igneous rocks. These rocks may form by pervasive intrusions or by partial melting of metamorphic rocks. Migmatites are common in the core of the Coast Plutonic Complex.

Milankovitch cycles: Cyclic changes in the earth's orbit that result in variations in absorption of the sun's energy. Milankovitch cycles may have caused changes in the earth's climate and ice ages.

mineral: Solid crystalline materials with a definite range of chemical composition and physical properties. Minerals are generally classified based on the anion or anionic group. Oxides (oxygen is the anion) and silicates (silica or SiO_2 is the anionic group) are important examples.

moraine: Piles and ridges of gravel and other sizes of rock fragments deposited by glacial ice. Some varieties are end, terminal, medial, and lateral.

mudstone: Rocks formed from solidification of fine-grained sediment or mud. Some varieties of mudstone have fissile layering and are known as shale. Other varieties have no obvious internal structure.

olivine: A yellow-green iron magnesium silicate mineral mostly found in silica-poor igneous rocks.

outwash complex: A shallow and wide plain of sediment derived from a glacier and then transported farther downhill by meltwater from the glacier.

paleoclimate: The earth's climatic conditions in the prehistoric past.

paleomagnetism: Study of magnetic field orientations preserved in rocks when they form. Information may include the direction to the north magnetic pole. This can be used to extract the latitude of the rock during formation.

peridotite: An intrusive igneous rock composed of olivine and pyroxene.

phyllite: A fine-grained metamorphic rock with a distinctive shiny luster that forms due to recrystallization of fine-grained sediments like mud into micas with subparallel alignment.

pillow texture: A structure found in volcanic rocks that comprises rounded or globular masses from a few centimeters to meters in size. These textures form when magma extrudes into water.

placer gold: Flakes and nuggets found in stream sediments and concentrated in pools from the hydraulic action of water on the high-density "heavy" gold.

plagioclase: A common mineral of the feldspar group composed of sodium, calcium, aluminum, silicon, and oxygen. This mineral is abundant in diorite, granite, and many metamorphic rocks.

plastic flow: A slow permanent deformation or flow of material. Ice under pressure within a glacier and hot rocks, generally deep in the earth, may flow like plastic.

plate boundary: The faulted boundaries between adjacent plates of the earth's crust; these boundaries are divergent (plates move apart), convergent (plates move toward one another and one is subducted if it is oceanic), or transform (plates slide horizontally past one another).

plate tectonics: The governing theory for global movement and interaction between rigid crustal plates near the earth's surface, and between these plates and the earth's interior.

pluton, plutonic: A body of or pertaining to magma that has crystallized into an igneous rock beneath the earth's surface.

pyrite: Cubic shiny yellow-gold mineral formed from iron sulfur. This mineral is often called fool's gold.

pyroxene: A group of iron, magnesium, and calcium silicate minerals found in igneous and metamorphic rocks.

quartz: An extremely common colorless to light-colored mineral composed of silicon and oxygen.

radioactive decay: Breakdown of atoms by emission of subatomic particles. Radioactivity produces atoms that differ from the original atom plus heat.

radiolaria: Small marine organisms that form silica shells. These open-water creatures have inhabited the seas since the Cambrian period of earth history.

relict: A feature that remains despite later processes. For example, early sedimentary or igneous textures that survive metamorphism are relict.

rhyolite: A volcanic igneous rock with high silica and low iron and magnesium content. These rocks are often erupted in an explosive manner from continental volcanoes.

schist: Metamorphic rock with a shiny finely layered texture caused by parallel alignment of micas.

sedimentary rock: Rocks that are formed from weathered material or sediment at the earth's surface. These rocks are composed of materials deposited as clasts or precipitated from water.

sedimentation: The process of sediment accumulation into sedimentary layers either in air or water.

seismograph: Instrument that is used to show the nature of vibrational energy transmitted from an earthquake or human activity.

sillimanite: A high-temperature and medium-pressure mineral composed of aluminum silicon and oxygen (Al_2SiO_5). Sillimanite is most commonly formed by metamorphism of mudstone.

slate: Metamorphic rock with a layered texture caused by parallel alignment of microscopic micas.

suture zone: The boundaries between two or more exotic terranes, characterized by folding and faulting.

tarn: Generally used to describe a small lake in an ice-carved mountainous region.

tectonic: Forces that cause and structures that result from movement of the earth's crust. This includes forces that drive plate tectonics, folds, faults, and mountain building.

till: Poorly sorted sediment that is directly deposited by ice.

tonalite: Also called trondjhemite, an intermediate to felsic plutonic igneous rock that generally has a white and dark green or black color (salt-and-pepper appearance) resulting from a mixture of light-colored feldspar, light gray quartz, and dark hornblende or biotite.

ultramafic rock: Iron- and magnesium-rich and silica-poor rocks. These rocks are found on Duke Island and other Alaska-type zoned ultramafic bodies. The sparsely vegetated red bluffs found at Red Bluff Bay on southeastern Baranof Island result from the nutrient-poor soils derived from weathering of this rock type.

U-shaped valley: A steep-sided, flat-bottomed valley carved by glacial ice.

volcano: Generally cone-shaped hills or mountains formed where liquid rock or magma reaches the earth's surface (for example, Mt. Edgecumbe near Sitka). Low viscosity magmas, like basalt, form broad "shield" volcanoes (for example, Mauna Loa); high viscosity magmas, like rhyolite and andesite, form steep "strato" volcanoes (for example, Mt. Rainier).

zircon: An important mineral composed of zircon, silicon, and oxygen ($ZrSiO_4$). Trace amounts of uranium and lead in zircon crystals can be used to determine the age of igneous and metamorphic rocks.

Bibliography

Arendt, A.A., K.A. Echelmeyer, W.D. Harrison, C.S. Lingle, and V.B. Valentine. "Rapid Wastage of Alaska Glaciers and Their Contribution to Rising Sea Level." *Science* 297 (2002): 382–386.

Barron, E.J. "Paleoclimatology." In *Understanding the Earth: A New Synthesis,* edited by G.C. Brown, C.J. Hawkesworth, and R.C.L. Wilson, 485–505. New York: Cambridge University Press, 1992.

Begét, J.E., and R.J. Motyka. "New Dates on Late Pleistocene Dacitic Tephra from the Mount Edgecumbe Volcanic Field, Southeastern Alaska." *Quaternary Research* 49 (1998): 123–125.

Berg, H.C., D.L. Jones, and D.H. Richter. "Gravina Nutzotin Belt—Tectonic Significance of an Upper Mesozoic Sedimentary and Volcanic Sequence in Southern and Southeastern Alaska." *U.S. Geological Survey Professional Paper* 800D (1972).

Bowditch, N. *The American Practical Navigator.* Publication 9. Bethesda, Maryland: Defense Mapping Agency Hydrographic/Topographic Center, 1995.

Brew, D.A. "Plate-Tectonic Setting of Glacier Bay National Park and Preserve and of Admiralty Island National Monument, Southeastern Alaska." In *Proceedings of the Third Glacier Bay Symposium,* edited by D.R. Engstrom, 1–5. Anchorage, Alaska: National Park Service, 1993.

Brew, D.A., and D. Grybeck. "Geology of the Tracy Arm–Fords Terror Wilderness Study Area and Vicinity, Alaska," in "Mineral Resources of the Tracy Arm–Fords Terror Wilderness Study Area and Vicinity," *U.S. Geological Survey Bulletin* 1525-A (1984).

Brew, D.A., G.R. Himmelberg, R.A. Loney, and A.B. Ford. "Distribution and Characteristics of Metamorphic Belts in the South-Eastern

Alaska Part of the North American Cordillera." *Journal of Metamorphic Geology* 10 (1992): 465–482.

Brew, D.A., R.B. Horner, and D.F. Barnes. "Bedrock-Geologic and Geophysical Research in Glacier Bay National Park and Preserve: Unique Opportunities of Local-to-Global Significance." In *Proceedings of the Third Glacier Bay Symposium,* edited by D.R. Engstrom, 5–10. Anchorage, Alaska: National Park Service, 1993.

Brew, D.A., A.T. Ovenshine, S.M. Karl, and S.J. Hunt. "Preliminary Reconnaissance Geologic Map of the Petersburg and Parts of the Port Alexander and Sumdum 1:250,000 Quadrangles, Southeastern Alaska." *U.S. Geological Survey Open File Report* 84-405 (1984).

Buddington, A.F., and T. Chapin. "Geology and Mineral Deposits of Southeastern Alaska." *U.S. Geological Survey Bulletin* 800 (1929).

Cohen, H.A., and N. Lundberg. "Detrital Record of the Gravina Arc, southeastern Alaska: Petrology and Provenance of Seymour Canal Formation Sandstones." *Geological Society of America Bulletin* 105 (1993): 1400–1414.

Coney, P.J., D.L. Jones, and J.W.H. Monger. "Cordilleran Suspect Terranes." *Nature* 288 (1980): 329–333.

Connor, C., and D. O'Haire. *Roadside Geology of Alaska.* Missoula, Montana: Mountain Press Publishing Company, 1988.

Crawford, M.L., W.A. Crawford, and G.E. Gehrels. "Terrane Assembly and Structural Relationships in the Eastern Prince Rupert Quadrangle, British Columbia," in "Tectonics of the Coast Mountains, Southeastern Alaska and British Columbia," edited by H.H. Stowell and W.C. McClelland, *Geological Society of America Special Paper* 343 (2000): 1–21.

Dauenhauer, N.M. and R. Dauenhauer, eds. *Haa Shuká, Our Ancestors: Tlingit Oral Narratives.* Sealaska Heritage Foundation, Juneau. Seattle and London: University of Washington Press, 1987.

Decker, J.E., and G. Plafker. "Correlation of Rocks in the Tarr Inlet Suture Zone with the Kelp Bay Group," in "The United States Geological Survey in Alaska: Accomplishments During 1980," edited by W.L. Coonrad, *U.S. Geological Survey Circular* 844 (1982): 119–123.

Douglass, S.L. and D.A. Brew. "Polymetamorphism in the Eastern Petersburg Quadrangle, Southeastern Alaska," in "The United States Geological Survey in Alaska: Accomplishments During

1984," edited by S. Bartsch-Winkler, *U.S. Geological Survey Circular* 967 (1985): 89–92.

Eberlein, G.D., and M. Churkin. "Paleozoic Stratigraphy in the Northwest Coastal Area of Prince of Wales Island, Southeastern Alaska." *U.S. Geological Survey Bulletin* 1284 (1970).

Emiliani, C. *Planet Earth: Cosmology, Geology, and the Evolution of Life and Environment.* Cambridge and New York: Cambridge University Press, 1992.

Engebretson, D.C., A. Cox, and R.G. Gordon. "Relative Motions Between Oceanic and Continental Plates in the Pacific Basin." *Geological Society of America Special Paper* 206 (1985).

EPICA Community Members. "Eight Glacial Cycles from an Antarctic Ice Core." *Nature* 429 (2004): 623–628.

Eppenbach, S. *Alaska's Southeast: Touring the Inside Passage.* 6th edition. Old Saybrook, Connecticut: Globe Pequot Press, 1994.

Fletcher, H.J., and J.T. Freymueller. "New Constraints on the Motion of the Fairweather Fault, Alaska, from GPS Observations," *Geophysical Research Letters* 30 (2003): 1139.

Gehrels, G.E. "Reconnaissance Geology and U-Pb Geochronology of the Western Flank of the Coast Mountains Between Juneau and Skagway, Southeastern Alaska," in "Tectonics of the Coast Mountains, Southeastern Alaska and British Columbia," edited by H.H. Stowell and W.C. McClelland, *Geological Society of America Special Paper* 343 (2000): 257–283.

Gehrels, G.E., and H.C. Berg. 1992. "Geologic Map of Southeastern Alaska." *U.S. Geological Survey Miscellaneous Investigations Series Map* I-1867 (1992).

Gehrels, G.E., W.C. McClelland, S.D. Samson, P.J. Patchett, and J.L. Jackson. "Ancient Continental Margin Assemblage in the Northern Coast Mountains, Southeast Alaska and Northwest Canada." *Geology* 18 (1990): 208–211.

Goldfarb, R.J., D.L. Leach, W.J. Pickthorn, and C.J. Paterson. "Origin of Lode-Gold Deposits of the Juneau Gold Belt, Southeastern Alaska." *Geology* 16 (1988): 440–443.

Gore, R. "Cascadia: Living on Fire." *National Geographic* 193 (1998): 6–37.

Haeberli, W., H. Bösch, K. Scherler, G. Strem, and C. Wallén. *World Glacial Inventory: Status 1988*. International Association of Hydrological Sciences, United Nations Environmental Program, and United Nations Educational, Scientific, and Cultural Organization, Nairobi: World Glacier Monitoring Service, 1989.

Hollister, L.S., and C.L. Andronicos. "A Candidate for the Baja British Columbia Fault System in the Coast Plutonic Complex." *GSA Today* 7, no. 11 (1988): 1–7.

Hope, A., III., *Traditional Tlingit Country and Tribes Map*. 3rd ed. Juneau: Tlingit Readers, 2000.

Howell, D.G. [original publication 1985]. "Terranes," in "Shaping the Earth: Tectonics of Continents and Oceans," edited by E. M. Moores, *Readings from Scientific American* (1990): 98–111.

Hudson, T., K. Dixon, and G. Plafker. "Regional Uplift in Southeastern Alaska," in "The United States Geological Survey in Alaska: Accomplishments During 1980," edited by W.L. Coonrad, *U.S. Geological Survey Circular* 844 (1982): 132–135.

Hudson, T., G. Plafker, and K. Dixon. "Horizontal Offset History of the Chatham Strait Fault," in "The United States Geological Survey in Alaska: Accomplishments During 1980," edited by W.L. Coonrad, *U.S. Geological Survey Circular* 844 (1982): 1328–1332.

Jones, D.L., A. Cox, P. Coney, and M. Beck [original publication 1982]. "The Growth of Western North America," in "Shaping the Earth—Tectonics of Continents and Oceans," edited by E.M. Moores, *Readings from Scientific American* (1990): 156–176.

Kearey, P., and F.J. Vine. *Global Tectonics*. Oxford: Blackwell Scientific Publications, 1990.

Larsen, C.F., R.J. Motyka, J.T. Freymueller, K.A. Echelmeyer, and E.R. Ivins. "Rapid Uplift of Southern Alaska Caused by Recent Ice Loss." *Geophysical Journal International* 158 (2004): 1113–1133.

Loney, R.A., D.A. Brew, L.J.P. Muffler, and J.S. Pomeroy. "Reconnaissance Geology of Chichagof, Baranof, and Kruzof Islands, Southeastern Alaska." *U.S. Geological Survey Professional Paper* 792 (1975).

Miller, L.D., H.H. Stowell, and G.E. Gehrels. "Progressive Deformation Associated with Mid-Cretaceous to Tertiary Contractional Tectonism in the Juneau Gold Belt, Coast Mountains, Southeastern Alaska," in "Tectonics of the Coast Mountains, Southeastern Alaska and British Columbia," edited by H.H. Stowell and W.C.

McClelland, *Geological Society of America Special Paper* 343 (2000): 193–212.

Monroe, J.S., and R. Wicander. *Physical Geology: Exploring the Earth.* Pacific Grove, California: Brooks/Cole, 2001.

Morozov, I.B., S.B. Smithson, J. Chen, and L.S. Hollister. "Generation of New Continental Crust and Terrane Accretion in Southeastern Alaska and Western British Columbia: Constraints from P- and S-Wave Wide-Angle Seismic Data (ACCRETE)." *Tectonophysics* 341 (2001): 49–67.

Moores, E.M., and R.J. Twiss. *Tectonics.* New York: W.H. Freeman and Co., 1995.

Nolan, M., R. Motyka, K. Echelmeyer, and D. Trabant. "Ice-Thickness Measurements of Taku Glacier, Alaska, and Their Relevance to Its Dynamics." *Journal of Glaciology* 41 (1996): 139.

O'Sullivan, P.B., G. Plafker, and J.M. Murphy. "Apatite Fission-Track Thermotectonic History of Crystalline Rocks in the Northern St. Elias Mountains, Alaska," in "The United States Geological Survey in Alaska," edited by J.A. Dumoulin and J.E. Gray, *U.S. Geological Survey Professional Paper* 1574 (1997): 283–293.

Ouderkirk, K.A. "Tertiary Mafic Dikes in the Northern Coast Mountains, British Columbia." B.A. thesis, Bryn Mawr College, 1982.

Pang, K.D., and K.K. Yau. "Ancient Observations Link Changes in Sun's Brightness and Earth's Climate," *EOS* 43 (2002): 481.

Press, F., and R. Siever. *Earth.* 4th ed. New York: W.H. Freeman and Company, 1986.

Romm, J. "New Forerunner for Continental Drift." *Nature* 367 (1994): 407–408.

Ruddiman, W.F. *Earth's Climate: Past and Future.* New York: W. H. Freeman and Company, 2001.

Schoenfeld, E., and L. Thompson. "Caves on Prince of Wales Island." *Juneau Empire,* 1998.

Sisson, V.B., S.M. Roeske, and T.L. Pavlis. "Geology of a Transpressional Orogen Developed During Ridge-Trench Interaction Along the North Pacific Margin." *Geological Society of America Special Paper* 371 (2003).

Skinner, B.J., and S.C. Porter. *Physical Geology.* New York: John Wiley and Sons, 1987.

132 BIBLIOGRAPHY

Spotila, J.A., J.T. Buscher, A.J. Meigs, and P.W. Reiners. "Long-Term Glacial Erosion of Active Mountain Belts: Example of the Chugach–St. Elias Range, Alaska." *Geology* 32 (2004): 501–504.

Stone, D., and B. Stone. *Hard Rock Gold: The Story of the Great Mines That Were the Heartbeat of Juneau, Alaska.* Seattle: Vanguard Press, 1983.

Stowell, H.H. "Sphalerite Geobarometry in Metamorphic Rocks and the Tectonic History of the Coast Ranges near Holkham Bay, Southeastern Alaska." Ph.D. dissertation, Princeton University, 1986.

——— "Silicate and Sulphide Thermobarometry of Low- to Medium-Grade Metamorphic Rocks from Holkham Bay, South-East Alaska." *Journal of Metamorphic Geology* 7 (1989): 343–358.

Stowell, H.H., and M.L. Crawford. "Metamorphic History of the Western Coast Mountains Orogen, Western British Columbia and Southeastern Alaska," in "Tectonics of the Coast Mountains, Southeastern Alaska and British Columbia," edited by H.H. Stowell and W.C. McClelland, *Geological Society of America Special Paper* 343 (2000): 257–283.

Stowell, H.H., N.L. Green, and R.J. Hooper. "Geochemistry and Tectonic Setting of Basaltic Volcanism, Northern Coast Mountains, Southeastern Alaska," in "Tectonics of the Coast Mountains, Southeastern Alaska and British Columbia," edited by H.H. Stowell and W.C. McClelland, *Geological Society of America Special Paper* 343 (2000): 235–255.

Stowell, H.H., and R.J. Hooper. "Structural Development of the Western Metamorphic Belt, Southeastern Alaska: Evidence from Holkham Bay." *Tectonics* 9 (1990): 391–407.

Stowell, H.H., and W.C. McClelland, eds. "Tectonics of the Coast Mountains, Southeastern Alaska and British Columbia." *Geological Society of America Special Paper* 343 (2000).

Stowell, H.H., T. Menard, and C.K. Ridgway. "Ca-Metasomatism and Chemical Zonation of Garnet in Contact Metamorphic Aureoles, Juneau Gold Belt, Southeastern Alaska." *Canadian Mineralogist* 34, no. 6 (1996): 1195–1209.

Stowell, H.H., D.L. Taylor, D.K. Tinkham, S.A. Goldberg, and K.A. Ouderkirk. "Contact Metamorphic P-T-t Paths from Sm-Nd Garnet Ages, Phase Equilibria Modeling, and Thermobarometry: Garnet

Ledge, Southeastern Alaska." *Journal of Metamorphic Geology* 19 (2001): 645–660.

Taylor, D.L. "Samarium-Neodymium Garnet Geochronology and Silicate Mineral Thermobarometry of Late Cretaceous Metamorphism, Garnet Ledge, Southeast Alaska." M.S. thesis, University of Alabama, 1997.

White, B.M., S.L. Seifert, B.W. Hitchcock, S. O'Neel, R.J. Motyka, and C.L. Connor. "1999 Results of Student Field Survey at LeConte Tidewater Glacier, Southeastern Alaska," in *Geological Society of America, Program with Abstracts* 31 (1999): 7.

Wood, D.J., H.H. Stowell, T.C. Onstott, and L.S. Hollister. "^{40}Ar/^{39}Ar Constraints on the Emplacement, Uplift, and Cooling of the Coast Plutonic Complex Sill, Southeastern Alaska." *Geological Society of America Bulletin* 103 (1991): 849–860.

Index

Page numbers in italics refer to illustrations